Two Manuals Combined into One Book

Combat and Operational Stress Control

And

Marine Scouting and Patrolling

Department of Defense

Pentagon
Publishing

Field Manual
No. 4-02.51 (8-51)

Headquarters
Department of the Army
Washington, DC, 6 July 2006

COMBAT AND OPERATIONAL STRESS CONTROL

Contents

Distribution Restriction: Approved for public release; distribution is unlimited.

***This publication supersedes FM 8-51, 29 September 1994.**

Figures

- Combat and operational stress control (COSC) and combat stress control (CSC).
- Behavioral health (BH) and mental health (MH).

Additionally, please note that the term "battle fatigue (BF)" that is used in AR 40-216 is being replaced with the term "combat stress reaction (CSR)." This change is due to DOD Directive (DODD) 6490.5, which specified that all Military Services use the term CSR for the purpose of joint interoperability. A DOD (Health Affairs) working group with the Services later added the term "operational stress reaction (OSR)" to further characterize stress reactions experienced by Soldiers. The only difference between a CSR and an OSR is that it takes place in a combat environment. To reduce confusion from this change in terminology, this manual will use combat and operational stress reaction (COSR) as opposed to "BF". The COSR (battle fatigue) "casualties" are Soldiers experiencing a stress reaction in combat or operational environment.

Introduction

In our own Soldiers and in the enemy combatants, control of stress is often the decisive difference between victory and defeat across the operational continuum. Battles and wars are won more by controlling the will to fight than by killing all of the enemy combatants. Uncontrolled combat stress causes erratic or harmful behaviors, impair mission performance, and may result in disaster and defeat of COSC preventive measures. The COSC preventive measures are aimed at minimizing maladaptive stress reactions while promoting adaptive stress reactions, such as loyalty, selflessness, and acts of bravery.

This manual provides doctrinal guidance for controlling excessive stress in combat and other operational environments. It identifies command and leadership responsibilities for COSC. It identifies COSC consultation, training, and education assistance available for units. This manual provides definitive guidance to BH personnel and CSC units for their COSC mission and for management of COSR and other behavioral disordered patients (BDPs). It identifies the requirements for COSC consultation, planning, coordination, rehearsal, and implementation of the COSC plan contained in the FHP annex of the operation order (OPORD).

Many stressors in a combat situation are due to deliberate enemy actions aimed at killing, wounding, or demoralizing our Soldiers and our allies. Other stressors are due to the operational environment. Some of these stressors can be avoided or counteracted by wise command actions. Still other stressors are due to our own calculated or miscalculated choice, accepted in order to exert greater stress on the enemy. Sound leadership works to keep stressors within tolerable limits and prepares the troops mentally and physically to endure them. Some of the most potent stressors can be due to personal organizational problems in the unit or on the home front. These, too, must be identified and, when possible, corrected or controlled. Unit needs assessments (UNAs) can help BH providers identify specific stressor in a unit and develop interventions to help unit personnel cope.

This manual identifies MH sections and medical units, CSC, that provides COSC support to units. It identifies the requirements for COSC assistance, as a unit transitions through the different phases of a deployment. Chaplains, commander, leader and COSC Soldier mentors provide assistance with after-action review (AAR) and traumatic events management (TEM).

Chapter 1

Combat and Operational Stress Control

SECTION I — INTERVENTION AND CONTROL FOR THE COMBAT AND OPERATIONAL STRESS THREAT

STRESS CONTROL

1-1. Combat stress includes all the physiological and emotional stresses encountered as a direct result of the dangers and mission demands of combat, see AR 40-216. Combat and operational stress control in the Army may be defined as programs developed and actions taken by military leadership to prevent, identify, and manage adverse COSRs in units. This program optimizes mission performance; conserves the fighting strength; and prevents or minimizes adverse effects of COSR on Soldiers and their physical, psychological, intellectual, and social health. Its goal is to return Soldiers to duty expeditiously. According to DODD 6490.2, COSC activities include routine screening of individuals when recruited; continued surveillance throughout military service, especially before, during, and after deployment; continual assessment and consultation with medical and other personnel from garrison to the battlefield.

1-2. Combat and operational stress control is the commander's responsibility at all levels. The commander is assisted with his responsibility for COSC by his staff, unit leaders, unit chaplain (Appendix A), and organic medical personnel. The commander may also receive assistance from organic COSC personnel at brigade and above, and from corps and above medical company/detachment CSC BH personnel. The key concern to combat commanders is to maximize the return-to-duty (RTD) rate of Soldiers who are temporarily impaired or incapacitated with stress-related conditions or diagnosed behavioral disorders.

1-3. The purpose of COSC is to promote Soldier and unit readiness by—

- Enhancing adaptive stress reactions.
- Preventing maladaptive stress reactions.
- Assisting Soldiers with controlling COSRs.
- Assisting Soldiers with behavioral disorders.

Note. The word control is used with combat and operational stress, rather than the word management, to emphasize the active steps that leaders, supporting BH personnel, and individual Soldiers must take to keep stress within an acceptable range.

COMBAT AND OPERATIONAL STRESS THREAT

1-4. Many stressors in a combat situation are due to deliberate enemy actions aimed at killing, wounding, or demoralizing our Soldiers and our allies. Other stressors are due to the natural environment, such as intense heat or cold, humidity, or poor air quality. Still others are due to leaders' own calculated or miscalculated choices (for example, decisions about unit strength, maneuver, the time of the attack, and plans for medical and logistical support). Sound leadership works to keep operational stressors within tolerable limits and prepares troops mentally and physically to endure them. In some cases however, excessive stress can affect both leaders' and Soldiers' decision-making and judgment, resulting in missed opportunities, or worse, in high casualties and/or failure to complete the mission. Finally, some of the most potent stressors are interpersonal in nature and can be due to conflict in the unit or on the home front. Extreme reactions to such stressors may involve harm to self (as in the hypothetical case of a Soldier that becomes suicidal on discovering that his wife wants a divorce) or to others (as in the case of a Soldier that impulsively fires a weapon at his noncommissioned officer (NCO) out of rage over perceived unfairness). These stressors must be identified and when possible, corrected or controlled. For more information on the control of combat/operational stressors and for details about specific leader actions to control stress, see FM 22-51. Also see FM 4-02 and FM 4-02.17 for additional information on the general and medical threat.

EFFECT OF STRESS

1-5. Focused stress is vital to survival and mission accomplishment. However, stress that is too intense, or prolonged, results in COSR that impairs their ability to function effectively. Some stressors contribute to misconduct that requires disciplinary action and may take the Soldier from duty for legal action and incarceration. In a broader context stress may cause battle and nonbattle injuries through inattention, clumsiness, and reckless behavior. These resultant injuries can include equipment losses and friendly fire incidents. Stress may increase disease rates by disrupting hygiene and protective measures, and impairing the body's immune defenses. Stress may progress to BH disorders, or suicidal behaviors and/or homicidal behaviors. Excessive stress in combat contributes to lapses in operational and tactical judgment and to missed opportunities that could increase the numbers of Soldiers injured over time.

MODEL FOR COMBAT AND OPERATIONAL STRESS CONTROL INTERVENTIONS

1-6. Combat and operational stress control assessments are performed during all phases of combat operations, stability and reconstruction operations, and support operations so the term COSC may refer to any of these different types of Army operations. The COSC assessments are performed at unit and individual level. They consider the range of variables according to a model, which recognizes that biological, psychological, and social factors influence each other. The COSC assessment reviews the interaction systematically to a depth appropriate to the need. The assessment identifies which variables can be modified to improve coping or outcome. Based on these assessments COSC personnel recommend courses of action (COA) to the commander. They identify and initiate COSC interventions to improve unit effectiveness and Soldier efficiency and well-being.

1-7. Provided in Figure 1-1 is a conceptual model of stress, its mitigating and aggravating factors and potential outcomes on Soldiers and families. This model can be helpful when designing COSC interventions to improve short-term and long-term outcomes.

combat team [BCT], regiment, or division) the branch (such as infantry, artillery, or military police [MP]) and beyond the branch to the US Army level.

- Unit cohesion is the binding force that keeps Soldiers together and performing the mission in spite of danger and death. Cohesion is a result of Soldiers knowing and trusting their peers and leaders and understanding their dependency on one another. It is achieved through personal bonding and a strong sense of responsibility toward the unit and its members. The ultimate adaptive stress reactions are acts of extreme courage and almost unbelievable strength. They may even involve deliberate self-sacrifice.

MALADAPTIVE STRESS REACTIONS

1-12. Combat and operational stress reaction and misconduct stress behaviors comprise the maladaptive stress reactions.

COMBAT AND OPERATIONAL STRESS REACTION

1-13. The Army uses the DOD-approved term/acronym COSR in official medical reports. This term can be applied to any stress reaction in the military unit environment. Many reactions look like symptoms of mental illness (such as panic, extreme anxiety, depression, hallucinations), but they are only transient reactions to the traumatic stress of combat and the cumulative stresses of military operations. Some individuals may have behavioral disorders that existed prior to deployment or disorders that were first present during deployment, and need BH intervention beyond the interventions for COSR.

MISCONDUCT STRESS BEHAVIORS

1-14. Examples of misconduct stress behaviors are listed in Figure 1-3. These range from minor breaches of unit orders or regulations to serious violations of the Uniform Code of Military Justice (UCMJ) and the Law of Land Warfare. Misconduct stress behaviors are most likely to occur in poorly trained undisciplined Soldiers, however good and heroic, under extreme combat stress may also engage in misconduct. Generally, misconduct stress behaviors—

- Range from minor breaches of unit orders or regulations to serious violations of the UCMJ and the Law of Land Warfare.
- May also become a major problem for highly cohesive and proud units. Such units may come to consider themselves entitled to special privileges and, as a result, some members may relieve tension unlawfully when they stand-down from their military operations. They may lapse into illegal revenge when a unit member is lost in combat. Stress control measures and sound leadership can prevent such misconduct stress behaviors, but once serious misconduct has occurred, Soldiers must be punished to prevent further erosion of discipline. Combat stress, even with heroic combat performance, cannot justify criminal misconduct.

OVERLAPPING STRESS BEHAVIORS

1-15. The distinctions among adaptive stress reactions, misconduct stress behaviors, and COSRs are not always clear. Indeed, the three categories of combat and operational stress behaviors may overlap. Soldiers with COSR may show misconduct stress behaviors and vice versa. Soldiers with adaptive stress reactions may also suffer from COSR. Finally, excellent combat Soldiers that have exhibited bravery and acts of heroism may also commit misconduct stress behaviors.

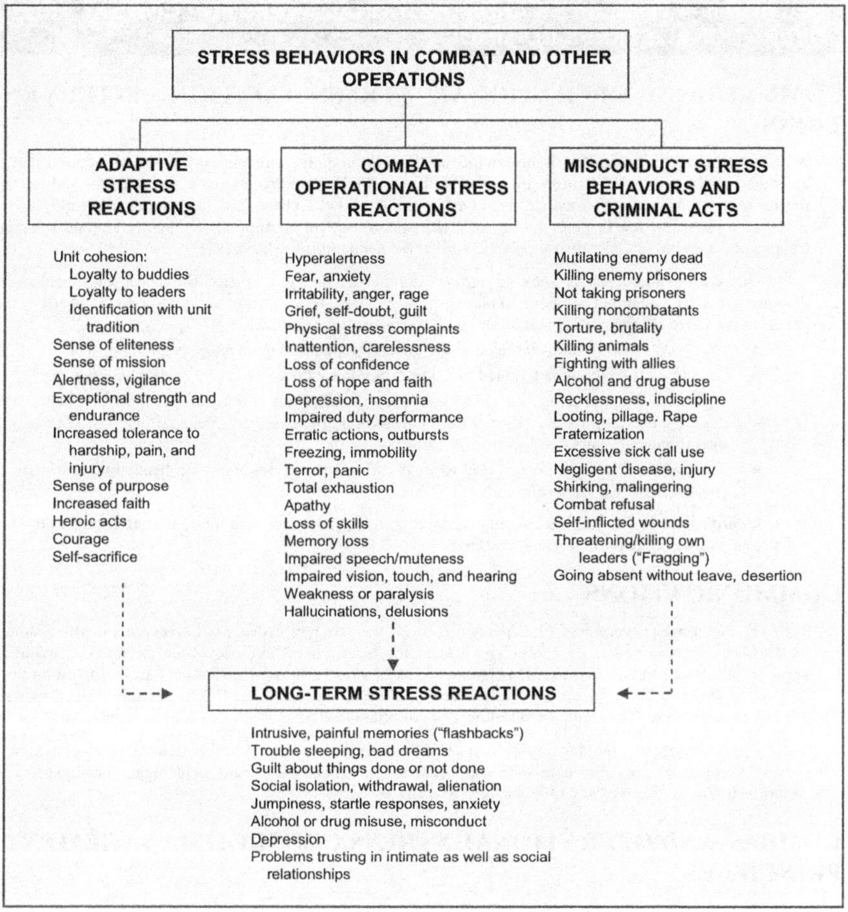

Figure 1-3. Stress behaviors in combat and other operations

1-16. It is common for stress reactions to persist or arise long after exposure to distressing events. When there is impairment in social and/or occupational functioning, a clinical assessment is warranted. Combat and operational stress control is important to sustain Army strength over the long-term and reduce the cost to society, the DOD, the Soldiers, and families.

SECTION II — PRINCIPLES AND FUNCTIONAL AREAS FOR COMBAT AND OPERATIONAL STRESS CONTROL

COMBAT AND OPERATIONAL STRESS CONTROL INTERVENTIONS

1-17. Soldier and unit readiness is best achieved through an active, prevention-focused orientation that is embodied in the principles. These principles apply to all COSC interventions or activities throughout the theater and are followed by medical personnel in all BH/COSC elements. Their application may differ based on a particular level of care and other factors pertaining to the mission, enemy, terrain and weather, troops and support available, time available, and civil considerations (METT-TC).

1-18. Preventive interventions seek to reduce the occurrence or severity of COSR and behavioral disorders, thereby sustaining Soldier and unit readiness. These interventions are tailored to the needs of the population. There are four categories of preventive interventions that include—

- Universal. Interventions targeted to the general population or an assigned AO.
- Selective. Interventions targeted to a unit or Soldier whose risk is higher than average.
- Indicated. Interventions targeted to Soldiers with COSR or indications of a potential behavioral disorder, and to units that show signs their mission effectiveness is being affected by combat and operational stressors.
- Treatment. Interventions targeted to treat and follow-up Soldiers with behavioral disorders to prevent their loss from duty.

1-19. Identify life- or function-threatening medical, surgical, or psychiatric condition as soon as possible and provide those patients emergency treatment.

COMMUNICATIONS

1-20. To maximize prevention, COSC personnel must maintain a high degree of involvement with Soldiers and leaders of supported units. Regular visits to the battalion/squadron level are essential to maintain appropriate accessibility. The COSC personnel must always be attuned to Soldiers around them and not rely on Soldiers to come to them. This means that when time permits, COSC personnel need to start conversations with as many supported Soldiers as possible.

1-21. Communication is the delivery system of COSC services and therefore is essential. Use all available means to communicate/coordinate with supported units, subordinate teams, and higher headquarters to ensure delivery of effective and timely COSC and BH services.

COMBAT AND OPERATIONAL STRESS CONTROL MANAGEMENT PRINCIPLES

1-22. The COSC management principles of brevity, immediacy, contact, expectancy, proximity, and simplicity (BICEPS). These principles apply to all COSC interventions or activities throughout the theater, and are followed by COSC personnel in all BH/COSC elements. These principles may be applied differently based on a particular level of care and other factors pertaining to METT-TC.

Note. The BICEPS principles are discussed in-depth in FM 6-22.5. This is a multiservice publication with the US Marine Corps (USMC). Due to differences in organizational structure the USMC defines the "C" in BICEPS to mean *centrality*, while the US Army defines it to mean *contact*, which will be used for the remainder of this manual.

BREVITY, IMMEDIACY, CONTACT, EXPECTANCY, PROXIMITY, AND SIMPLICITY

1-23. Using BICEPS is extremely important in the management of Soldiers with COSR and/or behavioral disorders.

Brevity

1-24. Initial rest and replenishment at COSC facilities located close to the Soldier's unit should last no more than 1 to 3 days (USMC and Navy is 3 to 4 days). Those requiring further treatment are moved to the next level of care. Since many require no further treatment, military commanders expect their Soldiers to RTD rapidly.

Immediacy

1-25. It is essential that COSC measures be initiated as soon as possible when operations permit. Intervention is provided as soon as symptoms appear.

Contact

1-26. The Soldier must be encouraged to continue to think of himself as a warfighter, rather than a *patient* or a *sick person*. The chain of command remains directly involved in the Soldier's recovery and RTD. The COSC team coordinates with the unit's leaders to learn whether the overstressed individual was a good performer prior to the COSR. Whenever possible, representatives of the unit or messages from the unit tell the Soldier that he is needed and wanted back. The COSC team coordinates with the unit leaders, through unit medical personnel or chaplains, any special advice on how to assure quick reintegration when the Soldier returns to his unit.

Expectancy

1-27. The individual is explicitly told that he is reacting normally to extreme stress and is expected to recover and return to full duty in a few hours or days. A military leader is extremely effective in this area of treatment. Of all the things said to a Soldier suffering from COSR the words of his small-unit leader have the greatest impact due to the positive bonding process that occurs. A simple statement from the small-unit leader to the Soldier that he is reacting normally to COSR and is expected back soon have positive impact. Small-unit leaders should tell Soldiers that their comrades need and expect them to return. When they do return, the unit treats them as every other Soldier and expects them to perform well.

Proximity

1-28. Soldiers requiring observation or care beyond the unit level are evacuated to facilities in close proximity to, but separate from the medical or surgical patients at the battalion aid station (BAS) or medical company nearest the Soldiers' unit. It is best to send Soldiers who cannot continue their mission and require more extensive intervention to a facility other than a hospital, unless no other alternative is possible. Combat and operational stress reactions are often more effectively managed in areas close to the Soldier's parent unit. On the noncontiguous battlefield characterized by rapid, frequent maneuver and continuous operations, COSC personnel must be innovative and flexible in designing interventions which maximize and maintain the Soldier's connection to his parent unit. See FM 6-22.5 for additional definition of proximity.

Simplicity

1-29. Indicates the need to use brief and straightforward methods to restore physical well-being and self-confidence.

1-30. The actions used for COSR control (commonly referred to as the 5 R's) involve the following actions:

- **R**eassure of normality.
- **R**est (respite from combat or break from the work).

Chapter 2

Behavioral Health and Combat and Operational Stress Control Elements in the Theater

SECTION I — MENTAL HEALTH SECTIONS

ASSIGNMENTS

2-1. Mental health sections are located in medical companies assigned to brigades, divisions, corps, and theater-level medical units.

ARMY OF EXCELLENCE

2-2. In the Army of Excellence (AOE) divisions, the division MH section (DMHS) is organic to the main support medical company (MSMC). In the armored cavalry regiments (ACRs) and separate brigades, a MH section is assigned to the forward support medical company (FSMC) of the forward support battalion (FSB) or a medical troop. In the corps and echelons above corps (EAC), the MH sections are located in the headquarters and headquarters detachment (HHD), area support medical battalion (ASMB) and in the area support medical companies (ASMCs). In those AOE divisions that are undergoing transformation, the MH sections are assigned to each of the brigade support medical companies (BSMCs) of the brigade support battalion (BSB) and in the ASMCs from the corps ASMB.

MODULAR FORCE

2-3. In the modular BCTs, Stryker BCTs (SBCTs) and the heavy BCTs (HBCTs) MH sections are assigned to the BSMC of the BSB. At echelons above brigade (EAB) and at the division, corps, or theater levels, the MH sections are assigned to the multifunctional medical battalion (MMB).

Note. Under the modular design, the MMB will replace the ASMB, the medical logistics battalion, and the medical evacuation battalion.

ORGANIZATIONS

2-4. The AOE divisions have only one MH section that is assigned to the MSMC of the main support battalion (MSB), which is located in the division support area (DSA). See Table 2-1 for a listing of personnel assigned to the division MH section. These personnel are deployed to the supported brigade as two-man teams comprised of one officer and one MH specialist.

Table 2-1. Division mental health staff

Division Psychiatrist (MAJ, Area of Concentration [AOC] 60W00, Medical Corps [MC])
Social Work Officer (CPT, AOC 73A67, Medical Service [MS] Corps)
Clinical Psychologist (CPT, AOC 73B67, MS)
Mental Health NCO (E7, E6, and E5, Military Occupational Specialty [MOS] 68X40/30/20)
Mental Health Specialist (Four)—(Two E4 and Two E3, MOS 68X10)

FUNCTIONS AND RESPONSIBILITIES OF ALL MENTAL HEALTH SECTIONS

2-5. All MH sections regardless of their organizational assignment are tasked with providing COSC for their supported units. In all of these units, COSC is accomplished through vigorous prevention, consultation, training, educational, and Soldier restoration programs. These programs are designed to provide BH expertise to unit leaders and Soldiers where they serve and sustain their mission focus and effectiveness under heavy and prolonged stress. The MH sections identify Soldiers with COSRs who need to be provided rest/Soldier restoration within or near their unit area for rapid RTD. These programs are designed to maximize the RTD rate of Soldiers who are either temporary impaired, have a diagnosed behavioral disorder, or have stress-related conditions. Also the prevention of posttraumatic stress disorders (PTSDs) is an important objective for brigades and EAB. The behavioral health officer (either a clinical psychologist or social work officer) and MH specialist are especially concerned with assisting and training of—

- Small-unit leaders.
- Unit ministry teams (UMTs) and staff chaplains.
- Battalion medical platoons.
- Patient-holding squad and treatment squad personnel of the medical company.

RESPONSIBILITIES

2-6. The MH section has a primary responsibility for assisting commanders with COSC by implementing the brigade combat mental fitness program. Also the MH section serves as a consultant to the commander, staff, and others involved with providing prevention and intervention services to unit Soldiers and their families. The MH section has staff responsibilities for assisting the brigade surgeon with establishing brigade policy and guidance for the prevention, diagnosis, treatment, management, and RTD of stress-related casualties. This is accomplished under the guidance and in close coordination with maneuver battalions and FSMC/BSMC physicians.

FUNCTIONS

2-7. The COSC functional areas were identified in Chapter 1. The functions of all MH sections are to support the nine functional areas except Soldier reconditioning.

UTILIZATION IN GARRISON

2-8. In garrison, BH personnel assigned to the division, brigade, or corps/EAC units continue to perform the same staff and outreach functions with supported units as they do in a field environment. An increase in the BH treatment functions may be possible as a result of consolidating BH care providers. The BH providers make available their consultation skills and clinical expertise to the Soldiers of supported units and their family readiness groups. Clinical care of family members and of Soldiers that require longer-term care beyond crisis intervention, brief treatment, and medication follow-up is the responsibility of the medical department activity (MEDDAC)/medical center (MEDCEN). The MH section personnel should focus their clinical work primarily on Soldiers with problems amenable to brief treatment.

2-9. Clinical services may be provided as part of a consolidated BH activity (as is usually established by a DMHS) or by augmenting an existing MEDDAC/MEDCEN BH staff.

2-10. The MH sections should strive to reduce referrals for BH treatment by working closely with unit leaders and chaplains to control organizational stress and rapidly identify and intervene with those Soldiers having BH disorders.

2-11. When the medical company or battalion deploys on training exercises, assigned BH personnel deploy with them to provide COSC training and support. In addition, they train to improve their own technical and tactical skills.

Note. In accordance with AR 40-216, clinical responsibilities in garrison must not interfere with participation in field and deployment exercises and maintenance of combat readiness.

MEDICAL DETACHMENT, COMBAT STRESS CONTROL

MISSION

2-12. The mission of the medical detachment, CSC (TOE 08463A000) is to provide COSC interventions and stress prevention activities to supported units in its AO. The staff augments division and brigade MH sections; provides direct support (DS) to combat brigades without organic BH officers; and provides area support in its AO. The medical detachment, CSC reconstitutes other brigade and division COSC assets. The medical detachment, CSC provides COSC interventions and activities to indigenous populations as directed in stability and reconstruction operations, humanitarian assistance, disaster relief, peace support operations, and detention facility operations. The medical detachment, CSC provides COSC interventions and activities to units in support of their readiness preparation and throughout their deployment cycle.

ASSIGNMENT

2-13. The Medical Reengineering Initiative (MRI) medical detachment, CSC (TOE 08463A000) is assigned to a corps medical brigade (MEDBDE) (TOE 08422A100), or other medical command and control (C2) element. Its teams may be further attached to a medical company, CSC (TOE 08467L000), an ASMB (TOE 08456A000), a BSMC/MSMC, or a combat support hospital (CSH). See Appendix B for the proposed future medical detachment, COSC.

EMPLOYMENT AND CAPABILITIES

2-14. The detachment headquarters is usually located in the DSA. Its teams disperse throughout the division's AO, and may extend support to corps if no CSC medical company is available. At TOE Level 1, the MRI medical detachment, CSC provides—

- A preventive section capable of dividing into four preventive teams. Each team provides mobile COSC interventions and activities.
- A fitness section capable of dividing into two fitness teams. Each fitness team provides mobile COSC interventions and activities. Each team is equipped to hold 40 Soldiers at the same time. With additional logistical support, each team can accommodate additional Soldiers under surge conditions.

STAFF RESPONSIBILITIES

2-15. The medical detachment, CSC assists the C2 headquarters (of the unit to which it is assigned or attached) regarding planning and coordination of COSC support, stress threat, mental and physical stressors, stress behaviors, principles of COSC, and implementation of COSC functional areas.

DEPENDENCY

2-16. This unit is dependent on—

- Appropriate elements of the corps for FHP, religious support, legal, finance, field feeding, personnel and administrative services support; laundry and clothing exchange; mortuary affairs support; and security of enemy prisoners of war (EPW), detainee, and US prisoner patients.
- The medical headquarters to which it is assigned/attached for FHP; medical administration, logistics, including MEDLOG; medical regulating of patients; evacuation; coordination for RTD; and unit level equipment and communications-electronics (CE) maintenance.

MOBILITY AND SECURITY

2-17. This unit is 100-percent mobile. It requires 100 percent of its organic personnel and equipment be transported in a single lift, using its organic vehicles. Upon relocation, Soldiers being held will require additional transportation. This unit is responsible for perimeter defense of its immediate operational area. However, it is dependent on appropriate elements of the corps for additional security, to include security of convoy operations. Personnel of the detachment are provided weapons for their personal defense and for the defense of their patients and/or held Soldiers.

ORGANIZATION

2-18. This 43-person MRI medical detachment (Figure 2-1) is organized into a headquarters section; a preventive section composed of four CSC preventive (CSCP) teams; and a fitness section consisting of two CSC fitness (CSCF) teams.

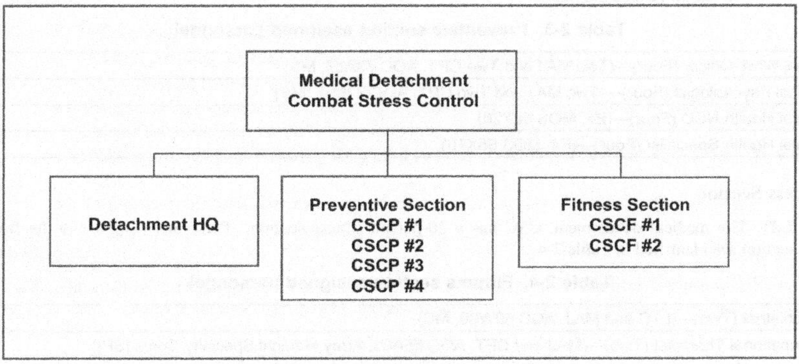

Figure 2-1. Medical Reengineering Initiative medical detachment, combat stress control (TOE 08463A000)

DETACHMENT HEADQUARTERS

2-19. The medical detachment, CSC headquarters provides C2 for the detachment. The headquarters element is responsible for planning, coordinating, and implementing COSC support for supported units. Personnel of the headquarters element provide maintenance, supply and service, and personnel administrative support. See Table 2-2 for a list of personnel assigned to the detachment headquarters.

Table 2-2. Detachment headquarters assigned personnel

Detachment Commander (LTC, AOC 60W00, MC)
Field Medical Assistant (1LT, AOC 70B67, MS)
Detachment Sergeant (E7, MOS 68X40)
Supply Sergeant (E5, MOS 92Y20)
Human Resource Specialist (E4, MOS 42A10)
Wheeled Vehicle Mechanic (E4, MOS 63B10)
Cook (E3, MOS 92G10)
Note. Detachment officers and NCOs from the preventive section and the fitness section may be assigned additional duties, which enhance the overall effectiveness of the headquarters section.

more of the medical company, CSC's eight preventive teams may locate at the BSMC when deployed in DS of SBCT, HBCT, and separate brigades or ACRs.

Fitness Teams

2-28. One or more of the four (ten-person) fitness teams may reinforce ASMCs which are deployed to locations throughout the corps and EAC. These teams provide a basis for COSC prevention and intervention. The teams may conduct Soldier restoration programs at the ASMCs, as required. These teams may also be deployed forward to provide temporary augmentation/reinforcement, as required.

2-29. Based on workload, one or more of the four fitness teams, plus one or more preventive teams locate with a theater hospital where they conduct Level IV COSC reconditioning programs, as required. A hospital located in the corps rear or out of the CZ is the best location to conduct the theater COSC reconditioning program. When deployed with a hospital, these teams provide mobile consultation in the vicinity of the hospital. These teams are also prepared to restrict reconditioning programs and deploy forward in support of higher priority missions on very short notice. These teams can also augment hospital BH ward services by staffing a temporary BH ward.

2-30. The medical company, CSC is divisible into four functionally emulative increments for split-based operations and stability and reconstruction operations as assigned. Nonstandard task elements for specific missions can be organized using any combination of the preventive section and fitness section personnel to meet specific BH needs. For stability and reconstruction operations, the minimum is an officer/NCO team to supplement a brigade COSC team or a preventive module/team of two officers, one NCO, and one enlisted. These modules may be augmented with personnel from the fitness section to add additional specialty expertise.

Headquarters Section

2-31. The headquarters section provides C2 and unit-level administrative and maintenance support to its subordinate sections when they are collocated with the company. The headquarters section may also provide assistance to detached elements by making site visits if the elements are within a feasible distance for ground transportation. The CSC medical company elements normally deploy with limited maintenance capability. When these COSC elements deploy, they are dependent on the supported units for patient accounting, transportation, food service, and field services. The personnel assigned to the headquarters section are identified in Table 2-5. See Appendix C for information on the AOE L-Series TOE medical company and medical detachment, CSC.

Table 2-5. Medical company, combat stress control, headquarters section personnel

Commander (LTC, AOC 60W00, MC)
Chaplain (CPT, AOC 56A00, Chaplain [CH])
Field Medical Assistant (CPT, AOC 70B67, MS)
First Sergeant (E8, MOS 68W50)
Mental Health NCO (E7, MOS 68X40)
Chemical Operations Specialist (E5, MOS 54B20)
Patient Administrative NCO (E5, MOS 68G20)
Supply Sergeant (E5, MOS 92Y20)
Power Generator Equipment Repairer (E4, MOS 52D10)
Wheeled Vehicle Mechanic (E4, MOS 63B10)
Administrative Specialist (E4, MOS 71L10)
Patient Administrative Specialist (Two)—(E4 and E3, MOS 68G10)
Armorer (E4, MOS 92Y10)
Cook (E3, MOS 92G10)
Note. Personnel from the headquarters section are deployed with teams or task-organized COSC element, as required.

SECTION II — OPERATIONAL STRESS ASSESSMENT TEAM IN A THEATER

MISSION OF OPERATIONAL STRESS ASSESSMENT TEAMS

2-32. The operational stress assessment team (OSAT) is task-organized when needed from personnel in the research institutes of United States Army Medical Research and Materiel Command (USAMRMC). The mission of this team is to deploy with BH research expertise, equipment and supplies to provide scientifically valid needs assessment of multiple units in the theater and macro analysis of COSC stressors, stress, and Soldier needs. The source is predominantly the Walter Reed Army Institute of Research, augmented when needed with research psychologists and enlisted technicians from the US Army Aeromedical Research Laboratory, the US Army Medical Research Institute of Chemical Defense, and the US Army Research Institute of Environmental Medicine. The OSATs are USAMRMC teams that are assembled entirely from personnel in table of distribution and allowances (TDA) units and are dependent on TOE units when they deploy to a theater.

ASSIGNMENT AND CAPABILITIES

2-33. The OSAT normally is used for augmenting the senior medical command in the theater. The OSAT has the capabilities to—

- Plan, coordinate, and conduct questionnaire and interview surveys.
- Establish a database and analyze scientific-quality data from their own surveys and from standardized unit assessments conducted by unit MH section and CSC units for their local supported units.
- Provide rapid reports of findings to the command that include recommendations for policy and for COSC actions and interventions.
- Design and validate survey instruments, including ones to address new and emerging stressors and issues in the theater.
- Use USAMRMC capability to archive and analyze data for comparison with other conflicts.
- Educate and mentor COSC and other personnel in survey methodology and provide them, as well as the theater commander, the wider perspective to use in their local UNA.

- Interface with the theater surveillance capabilities of preventive medicine (PVNTMED) and occupational and environmental medicine to address the BH aspects of chemical, biological, radiological and nuclear (CBRN) exposures and threats.

DEPENDENCY

2-34. The OSAT depends on the medical headquarters or the units with which it is working for administrative support. The OSAT depends on the theater (line) command for implementation of the approved surveys.

MOBILITY AND SECURITY

2-35. The OSAT depends on the medical and theater command and the units with which they work for transportation. The OSAT personnel are issued personal weapons and protective equipment, but depend on the units they are with for protection.

COMMUNICATIONS

2-36. The OSAT depends on the medical and theater command and the units with which they work for communications. With approval of theater command, they may be allowed USAMRMC-issued phones for satellite communication for mission-essential purposes.

ORGANIZATION

2-37. The typical OSAT has three officers and two enlisted Soldiers. The officer positions are listed as research psychologists (AOC 71H), but may be filled by clinically qualified BH officers with research training and experience. The enlisted are usually MH specialists (MOS 68X) with research training and experience. Staffing may also include personnel with information management specialties. The equipment is the required personal equipment of the team members, plus data collection and automated data processing equipment appropriate to the mission.

Chapter 3

Combat and Operational Stress Control Support Operations

SECTION I — COMBAT AND OPERATIONAL STRESS CONTROL PROFESSIONAL DISCIPLINES AND PROFESSIONAL CONSULTANTS

PROFESSIONAL DISCIPLINES

3-1. There are five BH professional disciplines and two enlisted specialties that serve in support of the COSC mission. The professional disciplines and enlisted specialties were identified in Chapter 2. The professional disciplines include social work, clinical psychology, psychiatry, occupational therapy (OT), and psychiatric/BH nursing; the enlisted specialties are in BH and OT. While much of the COSC knowledge base and most of the skills are shared by all the BH personnel, each discipline brings its own perspective from its professional training, skills that can only partially be familiarized to the others, and in some cases unique credentials to conduct specific assessments and treatments. The special perspectives and skill sets of the officer and enlisted members of the COSC team are summarized in the following subparagraphs.

OFFICER AREAS OF CONCENTRATION

Social Work Officer

3-2. The social work officer (AOC 73A67) evaluates the psychosocial systems affecting the Soldier, his family, and the unit; identifies and coordinates with military and civilian support agencies (including the Army Substance Abuse Program and the Family Advocacy Program [FAP]); assesses BH disorders; provides a range of counseling and psychotherapies. Installation MTFs are responsible for BH services that are beyond the scope of COSC functional areas such as long-term clinical care, inpatient services, and alcohol treatment programs. This officer provides proactive consultation, conducts individual and group counseling, supervises Soldier restoration/reconditioning, and coordinates RTD of recovered cases. He also provides staff advice and coordinates Army and civilian social services support.

3-3. When deployed as a member of a COSC preventive team or task-organized COSC element, the social work officer's duties include—

- Evaluating psychosocial (unit and family) functioning of Soldiers with COSR and misconduct stress behavior.
- Coordinating and ensuring the return of recovered COSR and neuropsychiatric (NP) Soldiers to duty and their reintegration into their original or new unit.
- Identifying and resolving organizational and social environmental factors which interfere with combat readiness.
- Coordinating support for Soldiers and their families through Army and civilian community support agencies, when possible.
- Apprising unit leaders, primary care physicians, and others health care providers of available social service resources.
- Providing consultation to supported unit commanders and to other BH/COSC personnel regarding problem cases.
- Counseling and providing therapy for Soldiers with psychological problems.
- Conducting and supervising unit needs assessment and TEM.

3-4. The doctoral level social work officer, as outlined in DODD 6490.1, may conduct clinical evaluations for imminent risk or dangerousness.

Clinical Psychologist

3-5. A clinical psychologist (AOC 73B67) diagnoses BH disorders and a range of psychotherapies including cognitive/mental approaches, assessment for safety evaluations, and psychometric assessment.

Psychiatrist

3-6. The psychiatrist (AOC 60W00) is a physician that is credentialed to diagnose and treat BDPs. He also conducts mental assessments, makes differential diagnosis, provides a range of psychotherapies and safety evaluations, and prescribes medication.

Psychiatric/Behavorial Health Nurse

3-7. This AN officer (AOC 66C007T) provides specialized nursing services for emotionally distressed individuals and promotes COSC within the MTF and the adjacent military community. He performs liaison and consultative functions to ensure continuity of patient care. The additional skill identifier of 7T indicates that advanced skills and competencies are required for this position and that the individual may act as a consultant in clinical nursing practices.

Occupational Therapist

3-8. The occupational therapist (AOC 65A00), plans, implements, and supervises preventive and restorative OT services to enhance the occupational performance of Soldiers. This officer evaluates Soldier's performance across the spectrum of occupational areas (such as activities of daily living, work, education, leisure, and social participation). Analyze jobs and job tasks for underlying performance requisites. Identify and evaluate mental and physical stressors and stress reactions and teach prevention, adaptive coping and psychosocial skills. Apply therapeutic media in both, individual and group settings to include environmental adaptation, the therapeutic use of self, consultation, education, and occupation (purposeful activity). Match Soldiers to therapeutic and/or vocational activities based on therapeutic need and functional ability.

ENLISTED MEDICAL OCCUPATIONAL SPECIALTIES

MENTAL HEALTH SPECIALIST

3-9. The MH specialist (MOS 68X30/20) provides the perspective of the enlisted Soldier, takes the initial history and conducts mental status evaluations for supervising officer, and administers psychological tests. The MH specialist also performs counseling and may have advanced specialty training (such as drug and alcohol counseling with higher ranked individuals).

OCCUPATIONAL THERAPY SPECIALIST

3-10. The OT specialist (MOS 68W30/20N3) provides the perspective of the enlisted Soldier, assists the supervising OT officer in evaluating a Soldier's occupational performance; conducts initial occupational performance history interviews and mental status evaluations; observes clients to gather data as part of task performance skill assessments; and implements OT interventions under the supervision of an OT officer.

CONSULTANTS

3-11. There are two levels of COSC consultants—the theater-level consultant and the subordinate commander consultant. The theater-level COSC consultant is normally the senior COSC officer designated by the theater surgeon and approved by the theater commander. The subordinate command COSC consultants are MH officers that function as the point of contact (POC) for one or more commanders.

These consultants oversee, coordinate, and conduct COSC interventions and activities for their respective commanders. They accomplish the following tasks appropriate to their level of operation:

- Direct and report AO-wide COSC needs assessments.
- Coordinate AO-wide implementation of COSC functional areas.
- Ensure COSC standards are established and consistently applied.
- Ensure compatibility between COSC standards and hospital BH standards (theater/AO COSC consultant).
- Define COSC operational requirements.
- Recommend the AO COSC evacuation policy (theater/AO COSC consultant).
- Prepare the COSC portion of the FHP estimate and operations plan.
- Review all command medical policies affecting COSC interventions and activities.
- Evaluate the quality of COSC interventions and activities rendered in the AO.
- Make regular AO COSC consultant site visits to COSC sections and units.
- Coordinate with other COSC consultants in adjoining AOs and in the primary medical support elements to ensure unity of effort.
- Oversee AO COSC training.
- Coordinate joint and combined COSC interventions and activities as necessary.
- Plan for future COSC operations.
- Consult with the command surgeon on all COSC matters.

SECTION II — DEPLOYMENT AND EMPLOYMENT OF MENTAL HEALTH SECTIONS

MENTAL HEALTH SECTIONS

3-12. The MH section assignments, staff, functions and responsibilities were discussed in Chapter 2. The information provided in this section pertains to deployment and employment and how these sections prepare for COSC support for their units in garrison, combat operations, and stability and reconstruction operations predeployment. The involvement of MH section with support units during field and other training exercises is important. It permits BH personnel to gain a familiarity with type of units supported, how they conduct operations, their mission, and the likely stressors associated with a particular type of unit. It also allows them to interact and become familiar with unit personnel and to gain some level of trust through their interactions. Combat and operational stress control personnel should be active participants in all training exercises. Unit predeployment training must be planned when the unit is alerted for its deployment but scheduled after cross-leveling and filling of all positions is completed. Training of BH personnel will include—

- Ensuring their active participation in the normal training cycle of the BCT/ACR to which assigned.
- Training the BH officers and specialists in all COSC functional areas that support their unit.
- Working with the supported units on a frequent basis so BH personnel may gain a familiarity with the unit's personnel, capabilities, and possible stressors associated with its mission.
- Providing predeployment train-up to include refresher and operation-specific training before deployment.

PLANNING

3-13. Since planning is normally performed at battalion level, MH sections are involved in developing a COSC estimate to support the unit plan. The COSC estimate of the situation is derived from mission analysis, to include considerations for COSR, BH disorders, substance abuse, misconduct behaviors, suicide, major stressors, troop populations to be supported, and COSC augmentation requirements or assets available.

COORDINATING

3-14. The BH personnel may be involved in coordinating COSC support with supporting and supported units, as appropriate. Brigade BH personnel must be known and trusted by the leaders of the units they support. Brigade BH personnel coordinate with other BH officer and enlisted personnel pertaining to the professional disciplines associated with conducting the COSC mission. They also coordinate with other key medical and nonmedical personnel (such as health care providers, surgeons, chaplains, UMTs that assist or may be involved with the COSC mission).

DEPLOYMENT OF MENTAL HEALTH SECTIONS

Early Deployment

3-15. The MH section should be deployed early into staging areas with the lead elements of the medical company and supported units.

Command and Control

3-16. The MH section falls under the C2 of the FSMC/BSMC and may receive mission guidance from the brigade surgeon through the FSMC/BSMC commander. Tasking for COSC operational support flows through the BSMC/FSMC headquarters. When BH personnel from the medical detachment/company, CSC are attached, the MH section ensures the timely flow of operations and situation awareness information to that element.

EMPLOYMENT OF THE MENTAL HEALTH SECTIONS

3-17. Employment of the MH section is based on the type of operations being conducted. The MH section normally operates from the FSMC/BSMC located in the brigade support area (BSA). When MH personnel are deployed forward, they are dependent on supported units for rations and logistical support.

MENTAL HEALTH FUNCTIONAL AREA CONSIDERATIONS

3-18. When determining COSC support for BCT units, the BH team should review each of the BH functional areas and determine which BH functions need to be implemented for each supported unit. Also, in developing the COSC plan, the BH team should consider—

- Conducting a UNA for each supported battalion and separate company supported.
- Establishing a consultation and education/training program for supported units to include—
  - Traumatic event and transition management.
  - The need for COSC triage, stabilization, and BH treatment should be addressed as they arise.
  - Soldier restoration, which can only be accomplished within and by the BSMC holding section or with augmentation by or referral to CSC unit personnel.
  - Reconditioning services that will need to be referred to nearest location where available.

SECTION III — SUPPORT OPERATIONS CONDUCTED BY THE MEDICAL UNIT, COMBAT STRESS CONTROL

PREDEPLOYMENT ACTIVITIES FOR ALL MEDICAL UNITS, COMBAT STRESS CONTROL

3-19. The predeployment activities identified in Section II for MH sections also apply for the units in Section III that include the ASMB or MMB and medical companies and detachments, CSC. These units must also keep themselves focused on applying the nine functional areas during the predeployment phase.

Note. The MMB is the C2 organization which will replace the ASMB, the medical logistics battalion, and the evacuation battalion.

Medical Detachment, Combat Stress Control Deployment

3-20. The medical detachment, CSC is deployed to provide COSC support whenever a division or two BCTs/regiments are deployed. It is assigned to a medical brigade or other medical headquarters for C2. Elements of the detachment may be further attached to FSMCs/BSMCs for operational control and logistical support while in support of the BCT.

Medical Company, Combat Stress Control Deployment

3-21. The primary mission of this unit is to provide COSC support throughout a corps AO. It is assigned to either a MEDCOM/MEDBDE and may support from two or up to five divisions. They assign teams to any operational area based on troop populations without BH support. They move teams rapidly to areas of surge workload or mass casualties.

Medical Force 2000 Verses the Medical Reengineering Initiative Detachments

3-22. The Medical Force 2000 (MF2K) and MRI CSC detachments conduct COSC support operations very similarly. However, in the MRI detachment, psychiatry assets are assigned to the fitness section and psychology assets are assigned forward to the prevention section as compared to the MF2K detachment, the manning strategy for psychology and psychiatry is reversed. The psychologists are assigned to the fitness section and psychiatrists are assigned to the prevention teams.

Medical Detachment, Combat Stress Control Support to a Division and Brigade Combat Teams During Combat Operations

3-23. When in support of an AOE division, a medical detachment, CSC is usually attached to the MSMC/BSMC and for a transformation division the medical detachment, CSC is attached to the MMB or MEDBDE. It can be attached to other battalion-sized units for support. It is under the operational control of the support battalion and MSMC/FSMC/BSMC and closely coordinates its augmentation support with the division surgeon and division psychiatrist. Habitual long-term relationships of medical detachments, CSC with specific divisions should be established so that COSC support is integrated as a normal function of the FHP mission. However, as an echelon above division (EAD) asset, the detachment (or its modular teams) may be task-organized and deployed to support other units or missions other than divisional units, as workloads require. The medical detachment, CSC depends on the units to which it is attached for administrative and logistics support. It may be reinforced and/or receive personnel replacements from a medical company, CSC or from another medical detachment, CSC.

Preventive Section Operations

3-24. Upon arrival to a theater, the detachment is initially attached to a higher medical headquarters for C2 and its three CSCP teams of four-person each are usually further attached. One CSCP is attached to each divisional MSMC prior to commencement of combat operations. This will allow the teams to linkup with and augment division COSC assets in support of the brigades.

- The deployment of each CSCP provides additional COSC personnel that increase COSC augmentation capabilities at the FSMC. A CSCP may routinely visit the BASs to provide consultation. The CSCP may deploy, as needed, to provide reconstitution support to units undergoing hasty or deliberate reorganization.
- In some operations, some or all of the CSCP personnel may either be attached to an ASMC or are deployed in support of a maneuver brigade or reconstitution site. It is important for all CSCP personnel to understand that they are responsible for all functional areas, and will conduct them as dictated by the mission.

Fitness Section Operations

3-25. The CSCF may locate with any Level II or above MTF in the AO. This section has tents and is equipped to operate a Soldier restoration center and can supervise Soldier restoration activities to include consultation, rest, relaxation, and treatment as necessary for speedy RTD within three days. The CSCF also provides triage and stabilization at the MSMC and consultation to other division units located in the DSA and division rear areas. All CSCF members are responsible for providing support for all COSC functional areas as dictated by the mission.

3-26. The CSCF staff may move within the AO to temporarily reinforce or reconstitute a CSCP at the brigade level, or to escort COSR casualties to the Soldier restoration center. The CSCF personnel give reconstitution support to attrited units, especially when the units return to the DSA. The CSCF supplements the division psychiatrist; brigade COSC teams, CSCPs, and chaplains and leaders in after-action debriefings. The CSCF helps integrate recovered Soldiers and new replacements into units during reorganization activities. When either the BSA or DSA is tactically too unstable to allow Soldier restoration operations, the CSCF may locate with an ASMC or a CSH and attempt to continue their support to the division from that location.

COMBAT STRESS CONTROL DETACHMENT IN CONTINGENCY OPERATIONS

3-27. For contingency operations, a task-organized CSC element from the CSC medical detachment may be deployed. A CSCP supporting a maneuver brigade may operate out of a central base of operations. The team may deploy forward to the BSA (base camps or fire bases), in response to—

- Anticipated battle.
- After-action debriefing requirements.
- Alcohol/drug problems in a unit.
- Incidents of misconduct stress behaviors.
- High incidence of COSRs.
- Unit rotation in or out of theater.

3-28. Several CSCF and CSCP from two or more CSC medical detachments may be consolidated under the command of the senior medical detachment, CSC commander to staff a central reconditioning program for the AO. This may also function as an alcohol/drug detoxification rehabilitation program. This may be referred to as a COSC center and will also provide consultation and treatment support to MP confinement facilities where misconduct stress behaviors may have led to incarceration. This could also include support for EPW and detainee confinement facilities.

STABILITY AND RECONSTRUCTION OPERATIONS

3-29. In stability and reconstruction operations, if the force deployed is smaller than a division, a medical detachment, CSC would not be required to provide COSC support. In such cases, the medical company, CSC or the medical detachment, CSC may be tasked with providing either a COSC team or a task-organized COSC element. The COSC team or the task-organized COSC element is attached to the supporting medical headquarters or to an MTF and conducts its mobile consultation mission.

MEDICAL UNITS, COMBAT STRESS CONTROL INTERFACE AND COORDINATION REQUIREMENTS

3-30. The medical units CSC must interface with its higher headquarters element and with the unit to which it is attached. The higher headquarters may be a MEDCOM, MEDBDE, or an ASMB/MMB.

- Interface between the CSC units and their higher headquarters. The medical unit CSC interfaces with its higher headquarters pertaining to its assigned mission. It provides estimates and has input to the operation plan (OPLAN). The unit receives its OPORD from the higher headquarters. Interface between the unit and higher headquarters staff elements will include the following subject areas—

- The COSC operations.
- Assignment or attachment of the medical detachment, CSC elements.
- Daily personnel and equipment status reports.
- Class VIII (medical supply) status and resupply requirements.
- Casualty feeder reports.
- Operation orders.
- Personnel replacement for the detachment.
- Medical intelligence information.
- Behavioral health/COSC consultation tasking and results.
- Maintenance requirements and requests.
- Replacement and reconstitution operations.
- Civil-military operations.
- Communications (signal operation instructions [SOI], access to message centers and nets, and transmission of COSC messages through medical, land, and other channels).
- Mass casualty plan.
- Road movement clearances.
- Tactical updates.
- Contingency operations.
- Return-to-duty and non-RTD procedures.
- Medical evacuation procedures (air and ground ambulances).
- Changes in locations of supported unit.

- Interface and coordination with unit to which attached. The headquarters of the unit to which a medical detachment, CSC is under operational control (OPCON) or attached is responsible for providing the administrative and logistical support requirements of the detachment. These requirements are normally identified in the attachment order. The higher headquarters will be identified in the attachment order and the medical detachment, CSC will be coordinated with this headquarters prior to deployment. The medical detachment, CSC must coordinate with the headquarters staff according to the tactical standing operating procedures (TSOP) of the unit of attachment. The staff shares information with the detachment commander or his representative pertaining to the threat, tactical situation, patient/COSR casualty status, and changes in FHP requirements. Coordination activities and subject area information exchange should include—
  - Command and control procedures.
  - Status of FSMCs/BSMCs and CSCPs.
  - Communications and SOI.
  - Operational support requirements.
  - Civil-military operations.
  - Soldier restoration operations.
  - Reinforcement and personnel replacement.
  - Road movement and clearances.
  - Casualty reporting and accountability.
  - Patient-holding procedures.
  - Force protection.
  - Convoy operations.
  - Geneva Conventions.
  - Detainees/EPW.
  - Improvised explosive devices (IEDs).
  - Religious and cultural considerations.

RECONDITIONING CENTERS

3-31. The medical company, CSC task-organizes CSC elements to staff separate small reconditioning centers in locations that are relatively secure. However, under some circumstances, the company may consolidate teams to establish a large reconditioning center, which supports two or three divisions. Reconditioning facilities normally locate near a CSH. See Chapter 12 for definitive information on reconditioning center operations.

Chapter 4

Unit Needs Assessment

SECTION I — PRINCIPLES AND TENETS OF UNIT NEEDS ASSESSMENT

GENERAL PRINCIPLES

4-1. The UNA is a systematic process for identifying the COSC needs of units. The UNA allows COSC personnel to identify priorities for interventions and activities and for allocating resources. The UNA is not a clinical screening to identify individuals who have or are at risk for BH disorders problems, but rather evaluates the needs of the Soldier population and leads to more effective preventive COSC activities and early interventions. The UNA allows COSC personnel to—

- Identify and describe specific areas of COSC need.
- Discover factors contributing to the needs.
- Provide an assessment of the BH training needs of Soldiers, leaders, UMT, and medical personnel within the unit.
- Develop plans to meet or improve the COSC needs of Soldiers and units through prevention and early intervention activities.

SCOPE

4-2. Unit needs assessments can be conducted at various command levels from small to large units. The COSC personnel at each level should conduct UNAs for their supported units. Larger-unit UNAs can include the composite findings and recommendations of one or more UNAs completed at subordinate levels. For example, a brigade UNA may include the results of several subordinate battalion UNAs (consolidated for confidentiality). Generally, UNAs are not conducted below the company level, though exceptional circumstances may dictate a platoon or lower UNA.

TENETS

4-3. Tenets of UNAs—

- Gain commander approval and support prior to conducting the UNA.
- Protect anonymity and ensure confidentiality of Soldiers and commanders. This includes the protection of unit identification from higher headquarters.
- Provide the commander with an unbiased assessment.
- Consider the social, political, and organizational factors of the environment.
- Ensure that information sources represent the entire unit.
- Select an assessment method that is consistent with the operational situation.
- Limit overgeneralizing the findings from one unit to another or from one time or situation to another.
- Recognize that UNAs provide population-level assessments of COSC needs, not clinical screening tools to identify individuals who may benefit from COSC interventions.
- Distinguish between what respondents report they need and what interventions are required.
- Ensure the UNA is planned and coordinated at a level commensurate with the complexity of the assessment and/or situation before starting.

SECTION II — FOCUS AND METHODS OF DETERMINING UNIT NEEDS ASSESSMENT

AREAS OF FOCUS FOR A UNIT NEEDS ASSESSMENT

4-4. A UNA involves the systematic assessment of numerous areas of Soldier and unit functioning. A typical UNA includes, but is not limited to, the following areas:

- Major stressors impacting the unit.
- Level of unit cohesion.
- Well-being of unit Soldiers.
- Soldier concern about home-front issues.
- Soldier knowledge and skill for controlling combat and operational stress.
- Soldier ideas for addressing COSC needs.
- Soldier knowledge of accessing COSC resources.
- Barriers and stigma that prevent Soldiers from accessing COSC services.
- Training needs of Soldiers, leaders, UMT, and medical personnel on topics of COSC importance (such as buddy aid, suicide awareness, or suicide prevention). See FM 6-22.5, FM 22-51, and US Army Training and Doctrine Command (TRADOC) Pamphlet 600-22 (available at: http://www.tradoc.army.mil/tpubs/pamndx.htm) for additional information on suicide awareness and prevention.

METHODS OF UNIT NEEDS ASSESSMENTS

4-5. The UNA takes advantage of all available information. Various methods can be used to assess general unit needs and to identify issues that differ among subgroups (for example, gender, rank, or race/ethnicity). The use of multiple assessment methods is recommended. These methods may include—

- Interviewing Soldiers to hear their perceptions and concerns. Information gathered during the casual conversation, although informal, may afford valuable anecdotal information (as well as developing trust and familiarity).
- Reviewing policy documents (standing operating procedures [SOPs]).
- Interviewing key unit personnel (chain of command, chaplain, and medical personnel).
- Conducting structured group interviews (focus groups or unit survey interviews).
- Administering standardized surveys and questionnaires (paper/pencil or web-based).
- Monitoring trend indicators (such as high rates of BH referrals, sick call, or misconduct, Soldier suicide, sexual assault, fratricide or disciplinary actions).
- Using multiple methods for gathering information when possible to ensure different viewpoints are considered.

Planning Considerations for Unit Needs Assessment

4-6. The UNA varies in complexity and formality depending on the purpose of the assessment and the needs of the supported unit commander. Many factors determine the complexity and formality of a UNA, which in turn influence its feasibility.

Complexity

4-7. Complexity equates to the requirements and cost to complete a UNA. Complexity is influenced by factors such as—

- The size and number of units to be assessed.
- Geographic dispersion of the units and time constraints.

Formality

4-8. Formality equates to the degree that scientific principles and methods are employed in the conduct of the UNA. An example of a UNA with low formality is one where the assessing team uses nonstructured group and individual interviews and perhaps brief questionnaires they have developed. The information is obtained from key individuals and a convenience sample of troops. That may be sufficient to quickly identify problems and make recommendations to resolve them. A UNA with high formality is one that uses professionally validated questionnaires and structured focus group interviews according to standard protocols. The data is collected from a scientifically selected sample of the larger population. The data is analyzed by standard analysis programs, so that statistically selected samples of valid comparisons can be made with other similarly sampled units in other geographical locations and across time. The OSAT expertise and assistance may be necessary for some UNA instruments and methods. In general, increasing formality increases the complexity of the UNA. The UNA requires a higher level of formality as the need for objective data and scientific precision increases (such as when results from multiple units are to be merged or compared with other UNAs).

Feasibility

4-9. Feasibility is the ability to accomplish a UNA with available resources. In developing a particular UNA, trade-offs are made to achieve an acceptable level of data quality (formality) for an affordable cost (complexity).

UNIT NEEDS ASSESSMENT PROCESS

4-10. All UNAs are conducted following a three phase plan.

PREASSESSMENT

4-11. Preassessment is an initial phase to obtain command support, determine target issues, and select appropriate methods to use.

ASSESSMENT

4-12. Assessment is the phase for gathering, integrating, and analyzing information to identify the COSC needs of the unit.

POSTASSESSMENT

4-13. The principal task of the postassessment phase is to determine the COAs to present to the commander which address the identified COSC needs. These findings are then linked to a plan of action.

WHEN TO CONDUCT UNIT NEEDS ASSESSMENTS

4-14. Unit needs assessments may be conducted at various times throughout the deployment cycle. A UNA may be conducted—

- Prior to initiation of COSC interventions and activities while COSC personnel and units establish their support relationships.
- To assess the effectiveness of COSC interventions and activities that are in the process of being conducted or that have been completed.
- At the request of supported commanders.
- To monitor trend indicators (see Paragraph 4-5).
- After serious traumatic events and significant unit transitions.
- To collect unit information for COSC planning.

4-15. Unit needs assessments should be conducted during all types of deployments, including stability and reconstruction operations, and combat operations.

Chapter 5

Consultation and Education

SECTION I — PRINCIPLES AND TENETS OF CONSULTATION AND EDUCATION

GENERAL PRINCIPLES

5-1. Combat and operational stress control consultation is defined as the transmission of information through an interactive relationship between the consultant and consultees. Education is used here in a broader sense in that it is the transmission of information by any means. Examples of consultation include providing COSC advice, coaching, training, and planning assistance. During this process, the consultant learns about the consultees and their needs, and tailors the interactions accordingly. Examples of education include distributing flyers, video and radio broadcasts, and news articles. The consultation and education functional area supports the other COSC functional areas.

CONSULTANTS

5-2. The term BH consultant is used in the general sense to describe any person performing COSC consultation or education. In addition to those theater or subordinate commander COSC consultants identified in Chapter 3, all BH/COSC personnel may serve as consultants at their level. Familiarization training with other BH disciplines enriches the ability to serve as a consultant.

CONSULTEES

5-3. Depending on the units in the area of support, broad types or categories of personnel may be consultees. These personnel are involved in recognition and control of stress as a result of their position or duty assignments. Consultation and education consultees may include—

- The command or unit surgeon and his staff.
- Staff chaplain and UMT.
- The senior commander and the senior NCO of a battalion, brigade, division, or corps.
- Staff officers and NCOs, including adjutant and personnel (S1/G1), intelligence (S2/G2), operations (S3/G3), civil-military affairs (G9), and the Judge Advocate General (JAG).
- Medical personnel such as PVNTMED teams that have missions that often complement the COSC mission.
- Company grade leaders, especially company commanders, executive officers, first sergeants (1SGs) platoon leaders, platoon sergeants, company NCOs, and Soldiers that are trained to be peer mentors.

SOLDIER-PEER MENTORS

5-4. Soldiers selected by their commanders may be trained to provide COSC help-in-place assistance for COSC information to peers. They may also serve as a POC between fellow Soldiers and the COSC and UMT teams. Selected Soldier-peer mentors with additional training could assist commanders with conducting COSC training in mission risk assessments.

TENETS FOR CONSULTATION AND EDUCATION

5-5. Consultation and education are ongoing processes that are performed across the deployment cycle and the continuum of operations. They may be provided in response to a specific request by command or recommendation of COSC personnel. Consultation and education may be provided during routine scheduled meetings such as a commander's weekly update or a leader professional development class.

5-6. Effective consultation is accomplished by the consultant's active outreach. Consultation is best conducted through recurring face-to-face contact, preferably at the consultee's location. Telephone and radio may be used to setup initial meetings and provide follow-up consultation. Consultation may be conducted one-on-one or in small groups where interaction is feasible. When necessary and feasible, audio or video teleconference may suffice. Active outreach supports the functional areas particularly COSC triage and traumatic event and transition management.

5-7. Successful consultation depends on the consultant's credibility, and the trust and familiarity established with consultees. In addition to sound professional knowledge base and clinical skills, the consultant must have military bearing and knowledge of the military (including the units, missions, vocabulary, acronyms, and skills involved). Rapport is enhanced by the demonstration of the consultant's genuine interest in the consultee and the unit. There are some situations when consultation and education are more effectively provided by the COSC officers, the NCOs, or the junior enlisted personnel.

SECTION II — CONSULTATION, EDUCATION, AND PLANNING

CONSULTATION PROCESS

5-8. The following six steps outline the process for most consultations:

- Initiate the process by introducing yourself and your capabilities to key leadership and get approval to continue.
- Assess the needs of consultees and formulate ways to address them.
- Present COA to consultee and define goals and feasibility of alternative actions.
- Implement approved COA.
- Evaluate outcomes or progress.
- Plan follow-up actions.

5-9. Consultation and education encompasses a broad range of topics, extending from prevention to treatment. The process of consultation and education assists the consultee with anticipation, identification, and control of stressors (environmental, physiological, cognitive, and emotional) and stress reactions (adaptive and maladaptive stress reactions). The following list of COSC functions, resources and their availability provides examples of consultation and education topics:

- Operation risk factors.
- Individual risk factors.
- Stress moderators.
- Unit cohesion, esprit de corps, and morale.
- Critical event management.
- Barriers to care and overcoming stigma.
- Operations planning.
- Integration of new personnel to unit.
- Trust in equipment and supporting units.
- Tough realistic training.
- Home-front stressors.
- Indicators of unit stress level.
- Combat and operational stress behaviors.
- Combat and operational stress reactions.

- Misconduct stress behaviors.
- Mental disorders.
- Suicidal and homicidal behaviors.
- Command referral processes.
- Scapegoating behaviors.
- Rumor control.
- Physical needs (such as sleep, nutrition, and hygiene).
- Social/emotional needs (such as morale, welfare, recreation push package).
- Communication with home.
- Reintegration of Soldiers into unit.
- Alcohol and substance abuse.
- Procedures for COSC triage.
- Care provider stress (such as compassion fatigue, vicarious trauma, moral dilemma).
- Long-term consequences of COSR and mental disorders.
- Medical evacuation policy and procedures.
- Leadership.
- Fear management.
- Rest and relaxation (R&R) considerations.

5-10. Additional details about some of these topics may be found in FM 22-51.

5-11. From theater staff level down to the MH sections, BH personnel are involved in the consultation and planning process for COSC support. Their support may be either through providing situational updates, developing staff estimates, or being directly involved and assisting with the development of the FHP plan. Planning starts with mission analysis. The mission analysis is the first step of the military decision making in abbreviated planning for a time-constrained situation, see FM 5-0. A part of mission analysis is based on the commander's intent and guidance. A medical battalion headquarters staff may receive addition instructions from the MEDBDE headquarters staff. During the planning process, command surgeons and medical planners (S3/G3) may seek information from the BH/COSC consultant or task medical company/detachment, CSC for COSC estimates.

MISSION ANALYSIS

5-12. The COSC planning is an essential part of every medical estimate and operations plan. Ensure that the command surgeon involves the COSC consultant in all medical planning. Mission analysis includes—

- Assessing COSC capabilities (organic and attached assets with current status and location).
- Gathering information from the commander's guidance regarding the upcoming operations. (See the S2 for intelligence, the S3 for scheme of maneuver and forces involved; the S1 for casualty estimates, personnel status issues, and units supported, and the logistics staff officer [S4/G4] for logistical considerations). Surgeons provide disease and nonbattle injuries (DNBI) estimates and status of medical units. Unit needs assessments, organic medical personnel, and UMT can provide additional information.
- Developing the COSC estimate based on the gathered information. The COSC estimate helps medical planners to anticipate demands on medical and COSC resources and to prepare accordingly. See FM 3-0, FM 5-0, and FM 8-55 for methods to develop the COSC estimate.
- Assessing limitations (specify reasons that BH assets are not available).
- Identifying specified, implied and essential COSC task in the FHP annex of the OPORD.
- Providing COSC COA.
- Presenting estimate and COA to command surgeon. Define goals and feasibility.
- Implementing approved COA. Consult with implementing COSC personnel about estimate and operational details.

- Evaluating outcomes or progress.
- Providing estimate and COA for fragmentary order (FRAGOs) and planning follow-up actions.

REHEARSAL

5-13. To achieve optimal synchronization, the FHP plan is rehearsed as an integral part of the combined arms plan at the combined arms rehearsal. Medical leaders provide input to the over all plan and develop the concept for the FHP plan. During the decision-making/orders process, they identify critical events and synchronize the FHP plans. In addition to medical locations on the CSS overlay, these plans indicate the triggers for FHP events. At the rehearsals, leaders practice their synchronized plans that include FHP.

5-14. The CSS/FHP annex of the OPORD that includes map overlays is the conclusion of the medical planning efforts; the rehearsal is the culmination of the preparation phase for an operation. The medical platoon leader has the responsibility for rehearsing FHP operations. Rehearsals are done to achieve a common understanding and a picture of how the plan will be implemented. The following leads to a successful rehearsal—

- All plans must be completed prior to the rehearsal.
- The FHP portion of the battalion rehearsals should focus on the events that are critical to mission accomplishment. A successful rehearsal ensures explicit understanding by subordinate medical personnel of their individual missions; how their missions relate to each other; and how each mission relates to the commander's plan. It is important for all medical units to understand the complete FHP concept.
- Rehearsing key FHP actions allows participants to become familiar with the operation and to visualize the "triggers" which identify the circumstances and timing for friendly actions. This visual impression helps them understand both their environment and their relationship to other units during the operation. The repetition of critical medical tasks during the rehearsal helps leaders remember the sequence of key actions within the operation and when they are executed.
- The OPORD is issued through effective troop leading procedures.

SECTION III — TRANSITION MANAGEMENT AND SUPPORT IN THE DEPLOYMENT CYCLE

SUPPORTING TRANSITION AND PHASES OF A DEPLOYMENT

5-15. Transition and phases of a deployment cycle may have different and sometimes unique stressors. These stressors may be subjects for consultation and education. Needs assessment may be necessary to redistribute COSC assets and change the means for delivering COSC activities and interventions.

5-16. The COSC personnel conduct transition (change of command) workshops, especially after relief of commanders. These workshops are normally requested by the incoming commander. The purpose of these workshops is to—

- Facilitate staff discussion of what the staff sees as the unit and staff's strong points and the areas needing more work.
- Provide the new commander the opportunity to discuss his leadership style and his expectations and set priorities for the staff.

5-17. The change of OPTEMP may include—

- Going from deployment to combat (offensive or defensive operations).
- Conducting battle handover in place by turning the battle over to another unit.
- Going from high intensity combat to stability and reconstruction operations.

PREDEPLOYMENT SUPPORT

5-18. Medical support in predeployment is mainly the responsibility of MEDDAC personnel or of a US Army Reserve (USAR) mobilization site augmentation unit (which has no BH personnel assigned).

Division and brigade COSC BH personnel are involved with their unit's Soldiers who are being treated for mental disorders, and with general COSC prevention of the unit members and families. The CSC units that are not deploying may be tasked to support predeployment COSC activities and interventions. Combat and operational stress control screening of individuals during predeployment may include—

- Secondary BH screening for Soldiers who are referred by primary care providers that flagged positive on DD Form 2795 (Pre-Deployment Health Assessment).
- Health screening of medical and BH records.
- Fitness for duty evaluations according to AR 40-501.
- Evaluations and recommendations to leadership about Soldiers who are fit for duty overall, but should not be deployed at this phase of the operation for BH reasons.

5-19. Other predeployment functions may be provided on request. These functions may include—

- Briefings, consultation and education on deployment cycle stressors and how to cope, and on specific stressors that Soldiers may encounter in the AO.
- Unit assessments at request of commanders.

Mid-Tour Rest and Recuperation or Emergency Leave

5-20. Leaders must be aware and alert for Soldiers who exhibit a need for COSC screening prior to mid-tour R&R. Other Soldiers may only require some education briefings and handout cards on handling their transition from a hostile CZ to R&R environments. In some cases, prior to emergency leave, some screening for domestic violence risk may also be necessary in any potential high-risk Soldier. This COSC screening is done for ensuring the safety of the Soldier and of others. One protocol has unit commanders referring those Soldiers that have requested leave for the purpose of addressing marital, legal, or other highly charged difficulties at home, to the BH officer. The BH officer explores the situation and the Soldier's reactions to it and checks for history with the rear detachment of the MH section, MEDDAC/MEDCEN and FAP. If the BH officer finds the Soldier to be at high risk, the commander can deny leave. Another option is to escort the Soldier to home base. At home base, he is kept by his unit's rear detachment until a meeting is held under safe conditions with the Soldier, BH, FAP personnel, his family, and the others involved in the difficulty. At this meeting the situation is defused or resolved before the Soldier is permitted to leave the rear detachment area.

Redeployment

5-21. For Soldiers, redeployment is the process of getting orders to redeploy, embarking towards the demobilization site, or deploying to another out-of-the-theater mission. Screening of these individual Soldiers may be a requirement and have established criteria that should be used to ensure appropriate screening is conducted. The theater or specific units may require that all primary care providers who perform screening interview, postdeployment health assessment have a designated BH consultant. Available COSC/BH personnel provide secondary one-on-one BH screening for Soldiers who are referred by primary care providers because they flagged positive on DD Form 2795. Some Soldiers may require brief treatment or referral may be required as a result of the screening.

End-of-Tour Stress Management

5-22. End-of-tour stress management is essential to reduce mental problems with Soldiers returning to their home station and/or families. This also promotes Soldiers seeking help early when problems occur. Mental health/COSC personnel consult with command on activities during redeployment and for postredeployment. Issues may include—

- Common mental disorders and potential misconduct may result when Soldiers are relieved of the stress and the focus of dangerous missions and are looking towards and preparing to go home.
- Memorial services for those Soldiers who died during the deployment.
- Recognition of outstanding performance and the equitable awarding of decorations.
- Closure ceremonies.
- Homecoming-reunion education briefing and training.

5-23. The end-of-tour debriefings (EOTDs) are like leader-led after-action debriefings, except that the time period covered extends from predeployment through to this point in redeployment. The EOTD should be conducted while the unit is still in the theater or at some place where transportation home pauses for one or more days. For each phase of the operation, the participants review what were the significant events and the problems or stressor that bothered the group (including critical events, if any). Equally important the group recalls what went well, how problems and stressors were overcome (at least by some members), and what positive memories and feelings they had and will take home.

Facilitators

5-24. The COSC or UMT facilitators should be invited to advise the leader and sit in if there are likely to be many negative memories and feelings that have disrupted or threaten unit cohesion. Advise leaders and troops on the importance of sustaining unit identity and contact with teammates beyond postdeployment and in the future.

Postdeployment

5-25. Postdeployment covers that period of time from embarking from theater, through scheduled postdeployment activities, and return to work after block leave. As with predeployment, medical support in postdeployment is mainly the responsibility of installation MTF, often augmented by a USAR mobilization site augmentation unit. Division and brigade COSC/BH personnel and CSC units that are themselves demobilizing as their units are deactivating will have limited involvement with the BH and COSC needs of fellow Soldiers. These personnel continue to provide, as needed, consultation to the command and to key personnel of the unit that they were assigned/attached. They consult with the installation MTF about BH and coordinate COSC needs that should be met. The CSC units that were not deployed may be tasked to support postdeployment COSC activities and interventions. The COSC personnel who are not demobilizing provide reintegration education. Screening of individuals postdeployment will continue and secondary screening will continue in postdeployment and be recorded on a DD Form 2796 (Post-Deployment Health Assessment). Each primary screener should have a designated BH consultant from the MEDCEN/MEDDAC. Secondary screening and treatment of referrals are provided by the installation MTF. Recent research with a returning division indicated that a secondary postdeployment screening from 90 and 120 days after returning from the theater may be necessary. This seems to be the time when persisting or delayed symptoms, problems, and perhaps impairment show need for intervention. All Soldiers should be advised about follow-up BH support resources. Soldiers that are leaving active service should also be informed about deployment-related entitlements and benefits.

Readiness Phase of the Deployment Cycle

5-26. Division/brigade MH sections and medical units, CSC, of Active Army and Reserve Component (RC) must use their training opportunities to the fullest. During these training exercises, they should provide COSC consultation, education, and training to supported units the same as they would in an actual deployment. Mental health personnel will perform many of their critical mission functions and should not just simulate them. The management of real COSR occurs in field and garrison/home-station settings. Triage and emergency stabilization may be required. Restoration may be appropriate in some field exercises. Even reconditioning can sometimes be provided as a multiday "course" at Active Army posts where CSC units are stationed. Further, TEM and transition management are also potential major roles for organic MH sections and medical units, CSC in peacetime. The RC as well as Active Army medical units, CSC have been used extensively to respond to traumatic events and to assist units that are in the premobilization, mobilization, or postdeployment phases of their deployment cycles.

Chapter 6
Traumatic Event Management

OVERVIEW OF TRAUMATIC EVENT MANAGEMENT

TAILORING TRAUMATIC EVENT MANAGEMENT

6-1. Traumatic event management blends other COSC functional areas to create a flexible set of interventions specifically focused on stress management for units and Soldiers following potentially traumatizing event (PTE). Like other functional areas, COSC providers must tailor COSC and TEM support to the needs of the unit and Soldier.

6-2. An event is considered *potentially traumatic* when it causes individuals or groups to experience intense feelings of terror, horror, helplessness, and/or hopelessness. Guilt, anger, sadness, and dislocation of *world view* or faith are potential emotional/cognitive responses to PTEs. Studies of Soldiers in Operation Iraqi Freedom (OIF) and Operations Enduring Freedom (OEF) have shown a correlation between exposure to combat experiences and BH disorders, most particularly acute stress disorder and PTSD. The following events should be monitored as PTEs for Soldiers and units:

- Heavy or continuous combat operations.
- Death of unit members due to enemy or friendly fire.
- Accidents.
- Serious injury.
- Suicide/homicide.
- Environmental devastation/human suffering.
- Significant home-front issues.
- Operations resulting in the death of civilians or combatants.

Scope of Traumatic Events Management

6-3. In this chapter, TEM is discussed as it applies to military units and personnel to support readiness. Traumatic events management can be adapted to nonmilitary groups or individuals brought together by a PTE. Traumatic events debriefing should be conceptualized as an ongoing process and not an acute intervention. For military units and personnel, TEM is active in all phases of the deployment cycle and across the continuum of military operations.

Traumatic Events Management Functional Mission Area

6-4. Successful TEM relies on a solid foundation of other COSC functional area activities such as UNA that may require the COSC provider to—

- Establish a credible working relationship with supported unit leadership.
- Understand the unit's needs via previously conducted UNAs.

6-5. In the absence of preexisting relationships and UNA, COSC providers should secure command support and recommend a UNA as the first step in TEM. Subsequent UNAs will clarify unit responses to other TEM interventions and to ongoing unit needs.

Unit Needs Assessments

6-6. When conducting a UNA after a PTE the COSC providers need to take under consideration the following—

- Ensure a timely arrival that does not disrupt unit operations but facilitates the UNA.

- They should not interrupt or intrude on the people who are attending to the acute crisis when arriving at the unit, unless asked. Presence without intruding will gain trust information and POC.
- Understanding that the UNA may be limited by the urgency of the unit's return to action, the difficulty of data collection, and having limited resources. Sufficient knowledge must be gained to tailor the interventions to the unit before initiating them. It is better to defer the intervention to the next opportunity if the unit must return to action immediately after replenishment and/or before necessary data is collected.
- Know that the UNA is a unit-level assessment and does not substitute for individual-level screenings or COSC triage. See Chapter 4 for a detailed discussion of UNA.

Consultation and Education

6-7. The COSC provider should conduct unit leader consultation and education activities prior to a PTE. Consultation and education topics should include—

- The impact of PTE on unit and Soldier readiness.
- Common PTEs for units and Soldiers.
- An overview of TEM.
- Components of TEM.
- Normal responses to PTEs.
- Triggers to refer Soldiers for BH evaluation.
- Development of SOPs for responding to PTEs when they occur.

6-8. Preemptive consultation and education prepare unit leaders to institute TEM interventions following a PTE. The 5 R's are a good model to build on. In acute TEM interventions, leaders should consider interventions that target—

- Safety, security, and survival.
- Food, hydration, clothing, and shelter.
- Sleep.
- Medication (replace medications destroyed or lost).
- Orientation of unit/Soldiers to developing situation.
- Restoration of communication with unit, dependents, friends, and community.
- Location should be at a secure site that provides protection from ongoing threats, toxins, and harm.

6-9. Consultation and education to Soldiers should emphasize normalizing the common reactions to trauma, improving their coping skills, enhancing self-care, facilitating recognition of significant problems, and increasing knowledge of and access to COSC services. Post-PTE UNAs guide further consultation and education efforts. Combat and operational stress control providers should be aware that leaders may not have experienced the PTE in person, but have experienced the PTE through their Soldiers and may require support.

6-10. For an in-depth discussion of consultation and education, refer to Chapter 5.

Combat and Operational Stress Control Triage

6-11. The COSC provider should be prepared to provide COSC triage in the aftermath of the PTE. Consultation and education is important to ensure appropriate and timely referrals for triage. The COSC provider can offer the following guidelines in referral of Soldiers for COSC triage:

- Persistent or worsening traumatic stress reactions (such as dissociation, panic, autonomic arousal, and cognitive impairment).
- Significant functional impairments (such as role/work relationships).
- Dangerousness (suicidal or violent ideation, plan, and/or intent).

- Severe psychiatric comorbidity (such as psychotic spectrum disorder, substance use disorder, or abuse).
- Maladaptive coping strategies (such as pattern of impulsivity or social withdrawal under stress).
- New or evolving psychosocial stressors.
- Poor social support.
- Failure to respond to acute supportive interventions.
- Exacerbation of preexisting psychiatric conditions.
- Soldier request for assessment.

Stabilization

6-12. The COSC provider should be prepared to provide or coordinate stabilization services following the PTE. Pre-PTE coordination with medical unit personnel promotes safe management.

Soldier Restoration

6-13. The measures below are applicable to Soldiers with COSR following a PTE. The COSC provider should be familiar with the 5 R's and with BICEPS. In keeping with restorative efforts, the COSC provider focuses on the following measures through leadership consultation, Soldier education, and/or direct management:

- Minimizing exposure of Soldiers with COSR to further PTE.
- Reducing physiological arousal.
- Mobilizing support for those who are most distressed.
- Providing information and fostering communication and education.
- Using effective risk communication techniques.
- Proving assurance/reassurance.
- Mitigating fear and anxiety.
- Encouraging sleep hygiene.
- Reestablishing routines.
- Promoting exercise and nutrition.
- Encouraging self-paced emotional ventilation.
- Discouraging use of alcohol/substances.

Behavioral Health Treatment

6-14. Given the correlation between PTE and development of traumatic stress disorders the COSC provider must be familiar with the best current practices for evaluation and treatment. The Veterans Health Administration (VHA)/DOD Clinical Practice Guidelines website (http://www.oqp.med.va.gov/cpg/cpg.htm) offers clinicians evidence-based assessment and treatment algorithms for acute stress disorder, PTSD, and many other BH disorders.

6-15. In recent years, the use of early interventions in response to PTE has come under critical review. The focus of much of this debate is on the use of psychological debriefing (PD) and more specifically, the critical incident stress debriefings provided to individuals or groups exposed to PTEs. Current research suggests that PD/critical incident stress debriefings can be harmful to participants, while failing to reduce traumatic stress reactions or to prevent the progression to PTSD. As PD/critical incident stress debriefings undergo definitive study, COSC providers should adhere to current evidence-based best practices.

6-16. In accordance with the VHA/DOD Clinical Practice Guidelines (http://www.oqp.med.va.gov/cpg/cpg.htm) for PTSD, COSC providers should—

- Consider alternative methods to PDs for individuals affected by PTEs.
- Avoid PD as a means to reduce acute posttraumatic distress (acute stress reaction or acute stress disorder) or to slow progression to PTSD.

- Understand there is insufficient evidence to recommend for or against conducting structured group debriefings.
- Be aware that compulsory repetition of traumatic experiences in a group may be counter-productive.
- Consider group debriefings with preexisting groups (such as teams, units, emergency medical treatment [EMT] teams, coworkers, family members) may assist with group cohesion, morale, and other important variables that have not been demonstrated empirically.
- Emphasize that group participation must be voluntary.

LEADER-LED AFTER-ACTION DEBRIEFING

6-17. A leader-led after-action debriefing is lead by a platoon, squad, or team leader and is not normally conducted above platoon level. The leader-led after-action debriefing should be conducted after all missions especially when the maneuvers did not go according to plan. A leader-led after-action debriefing may even be sufficient for PTEs involving injury or death. For the leader to conduct a PD, his personnel should have received previous PDs; and normally provide peer support and validation for showing and talking about their emotional reactions during and after the debriefing. The best time to conduct this debriefing is as soon as is feasible after the team/squad/platoon has returned to a relatively safe place and members have replenished bodily needs and are no longer in a high state of arousal. Usually a well-conducted leader-led after-action debriefing is the best option to manage PTEs during a mission. The exception to this type of debriefing is when the event evoked reactions that seriously threaten unit cohesion and/or have a high likelihood of arousing disruptive behavior and emotions. In these situations the leader should ask himself the following—

- Should I conduct the debriefing?
- Should a trained facilitator be present?
- Should a request for COSC TEM be submitted for his team/squad/platoon?

CONDUCTING A LEADER-LED AFTER-ACTION DEBRIEFING

6-18. These debriefings require the leader to extend the lessons-learned orientation of the standard AAR. He uses the event reconstruction approach or has the individuals present their own roles and perceptions of the event, whichever best fits the situation and time available. See Training Circular (TC) 25-20 for definitive information on AARs. When individuals express or show emotions, the leader and the teammates recognize and normalize them; they agree to talk about them later and support the distressed Soldier through personal interactions. The group then returns to determining the facts. Lessons-learned discussion is deferred until all the facts are laid out. See FM 22-51 for additional information. The leader may provide education about controlling likely reactions or referral information at the end, depending on his knowledge and experience.

6-19. When a PTE is likely to create individual or collective guilt, distrust, or anger, the unit leader should be encouraged to request COSC assistance. Either a COSC or a UMT person trained in TEM sits in with the leader-led debriefing as a familiar and trusted friend of the unit. The COSC or UMT facilitator helps the unit/team leader rehearse and mentors the leaders on the debriefing process. During the PD, the facilitator can ask questions of the group to clarify the facts and steer the process away from divisive anger, blaming, and scapegoating toward a positive, cohesion-restoring outcome. This method is halfway between a simple leader-led after-action debriefing and a PD and is referred to as a *facilitated leader-led after-action debriefing*. The leader conducting the debriefing must be attentive to identify individuals needing COSC follow-up.

6-20. Leaders in positions above platoon level also have a role in leader-led after-action debriefing. Company commanders and 1SGs may conduct leader-led after-action debriefing with their subordinate leaders. Battalion commanders and higher may conduct leader-led after-action debriefings with their staff after distressing actions and may include subordinate leaders when time allows bringing them together.

COOL DOWN MEETINGS

6-21. A cool down meeting is referred to as an immediate, short meeting when a team or larger unit/group returns from the battlefield or other missions. These cool down meetings are held after heavy/intense battles with the enemy or a shift in the mission has occurred which is highly arousing and/or distressing. This is especially important after critical events. The cool down meeting is an informal event and occurs before the participants fully replenish their bodily needs and precedes any other activities including more COSC interventions, or return to the mission.

6-22. Personnel who coordinate and wait for the return of the unit that has been in a heavy intense battle, include leaders or supporting officers or NCOs from the command, UMT, and COSC providers. (In domestic support operations the same personnel identified above or other trained personnel from governmental or nongovernmental organizations such as the Red Cross may be waiting on the return of the unit after a PTE. These personnel may be present at the cool down meeting.)

COMPONENTS OF A COOL DOWN MEETING

6-23. Components of a cool down meeting may include—

- Assembling all of the unit personnel at a safe and relatively comfortable location for a brief period of time (about 15 minutes).
- Receiving or sharing nonstimulating beverages and convenient food (*comfort* foods if available).
- Providing personnel the opportunity to talk among themselves.
- Giving recognition and praise for the difficult mission they have completed.
- Providing information to unit personnel on where and how they will rest and replenish.
- Previewing the immediate agenda for the unit on what will happen after the cool down meeting including plans for further debriefing and/or other available stress control or morale/welfare intervention.
- Providing announcements pertaining to further preparations and expected time of return to the mission.

6-24. The COSC personnel may have a role in cool down meetings. In consultation and education, COSC consultants emphasize the value of cool down meetings and the simplicity of the components, which are easily neglected in crisis situations. When feasible, they unobtrusively attend the cool down meeting, showing "presence" while learning of the event, getting familiar with the key people, and observing anyone showing signs of distress and being available to them. If requested, they may give a very brief introduction and review of normal stress reactions that unit personnel may have in the next few hours.

Chapter 7

Combat and Operational Stress Control Support for Units Undergoing Reconstitution

UNIT RECONSTITUTION SUPPORT

7-1. Unit reconstitution support is defined in FM 100-9 as extraordinary actions that commanders plan and implement to restore units to a desired level of combat effectiveness commensurate with mission requirements and available resources. Besides normal support actions, reconstitution may include—

- Removing the unit from combat.
- Assisting the unit with external assets.
- Reestablishing the chain of command.
- Training the unit for future operations.
- Reestablishing unit cohesion.

7-2. When tasked as part of a reconstitution task force, COSC personnel are responsible for providing units with Soldier restoration, performing the COSC functions of the UNA, and providing consultation and education, as required. They provide triage, stabilization, Soldier restorations, and short time-constrained COSC treatment, when needed. Reconstitution is a time-constrained process, but Soldier restoration may be provided at the reconstitution site if several days are available. The focus of consultation and education efforts include—

- Rebuilding unit cohesion.
- Integrating new Soldiers or groups of Soldiers into the unit.
- Facilitating assumptions of command by replaced leaders.
- Facilitating the building of Soldiers' confidence in their leaders and themselves.
- Mentoring unit leaders in leader-led after-action debriefing process.
- Advising on COSC aspects of bereavement memorial services and communication with the family support group and unit families. The COSC personnel also facilitate leader-led after-action debriefings and lead or serve as observers in TEM interventions.

RECONSTITUTION PROCESS

7-3. Reconstitution of units transcends normal day-to-day force sustainment actions. Reconstitution is a total process. Its major elements are reorganization, assessment, and regeneration, in that order. All COSC personnel should be thoroughly familiar with FM 100-9.

Reorganization

7-4. Unit reorganization primarily involves a shifting of internal resources and is accomplished as either immediate or deliberate reorganization. The commander of the attrited unit decides to reorganize when required and may consider the following—

- Immediate reorganization is the quick and usually temporary restoring of degraded units to minimum levels of effectiveness. Normally, the commander implements immediate reorganization at his combat location or as close to that site as possible to meet near-term needs. The COSC personnel provide consultation with key POC in the unit (leaders, UMTs, and trauma/health care specialist, usually by telecommunication).
- Deliberate reorganization is done to restore a unit to the specified degree of combat effectiveness. Usually, more time and resources are available further to the rear. Procedures are similar to immediate reorganization except that some personnel and weapons system

replacement resources may be available, equipment repair is more intensive, and more extensive cross- leveling is possible. When used in reorganization, cross-leveling involves the movement of personnel and/or equipment between units to achieve equalization. The process is accomplished while maintaining or restoring the combat effectiveness of the units involved.

- The role of COSC personnel in deliberate reorganization may require a COSC team to deploy to the reorganization site. They assess the stressors and stress reactions and advise the commanders on supportive actions, (such as those for the reorganization of small unit-level elements), assists command with transitions and integration of new replacements.
- When the reorganization involves battalion size or larger units the process becomes more like the coordinated operation described for unit regeneration, provided below. It is likely to draw upon fitness team expertise and Soldier restoration assets from rearward as well as on the forward deployed teams.

REGENERATION

ASSESSMENT PHASES

7-5. The defining characteristic of regeneration is the massive infusion of personnel, equipment, and assistance under the directions of higher headquarters. Assessment measures a unit's capability to perform its mission. It occurs in two phases. The unit commander conducts the first phase. He continually assesses his unit before, during, and after operations. If he determines it is no longer mission-capable even after reorganization, he notifies his commander. The higher headquarters either changes the mission of the unit to match its degraded capability or removes it from combat. External elements may also have to assess the unit after it disengages. This is the second phase. These elements do a more thorough evaluation to determine regeneration needs. They also consider the resources available.

7-6. The second phase of assessment begins with an initial survey by a team sent by the higher headquarters. This team determines the status and needs of the attrited and exhausted unit as it moves to the regeneration site. Some of the key issues in estimating the COSC needs of the unit include the—

- Percentage and nature of casualties.
- Duration of operations and environmental exposure.
- Status of hydration, nutrition, and sleep.
- Loss and current effectiveness of leaders.
- Attitudes, perceptions, and level of confidence of unit survivors.
- Traumatic events.

7-7. The COSC consultants at every level must provide consultation to all command surgeons on the importance of including BH/COSC personnel in all reconstitution operations. The initial evaluation team should include a subordinate commander COSC consultant. Task-organized CSC teams normally deploy to the unit during the second phase to provide UNA, consultation, and other COSC activities and interventions. Unit needs assessment with high formality may be requested by the higher headquarters for decision-making purposes. These assessments are feasible for selected units, with prior coordination, as much of the logistical complexity is reduced by the orchestrated planning at higher command echelons.

Conducting Regeneration

7-8. Regeneration involves rebuilding a unit through large-scale replacement of personnel, equipment, and supplies. If required, it includes internally reorganizing; reestablishing or replacing essential C2 and conducting mission-essential training for the reconstituted unit. Regeneration is required when heavy losses of personnel and equipment leave a unit combat-ineffective and unable to continue its mission. The commander controlling assets to conduct regeneration decides whether to use resources for this purpose. Regeneration has two variations: incremental regeneration or whole-unit regeneration. Incremental regeneration is the massive infusion of individual personnel replacements and single items of equipment into the surviving unit elements. Whole-unit regeneration is the replacement of whole units or definable subelements, such as squads, crews, and teams. The S1/G1, S4/G4, and medical staffs coordinate the dispatch of the regeneration task-force teams. These teams occupy the reconstitution site before arrival of the exhausted unit. The COSC assets that perform the indicated initial assessment are needed as part of this task force. The reconstitution task force guides each element of the arriving units into its designated areas. The regeneration task force provides for the immediate needs of the survivors. This should include personal gear and equipment to replace lost or damaged items, food services, personnel services, maintenance teams, and medical teams to provide sick call services while organic medical personnel rest. Replacement personnel are sent to the reconstitution site. The COSC personnel assist with their assimilation into the regenerated unit.

RECONSTITUTION RESOURCE REQUIREMENTS FOR COMBAT AND OPERATIONAL STRESS CONTROL

7-9. Factors which influence the resource requirements for COSC in reconstitution (deliberate reorganization and regeneration) include the size of the unit, number of subunits which have suffered heavy casualties, the extent of emotional trauma, and time available. The focus of COSC for reconstitution support was provided in Paragraph 7-2. A guideline is provided in Table 7-1 for COSC personnel requirements. This staffing should not overly rely on organic COSC staff that will be in need of rest.

Table 7-1. Reconstitution operations guideline for combat and operational stress control personnel requirements

Size of Unit	*Personnel Required*
Company/Troop	2 to 4
Battalion/Squadron	6 to 12
BCT/Brigade Regiment	12 to 30
Division	30 to 60

Chapter 8

Combat and Operational Stress Control Triage

SECTION I — THE TRIAGE PROCESS FOR COMBAT AND OPERATIONAL STRESS CONTROL

SOLDIER TRIAGE

8-1. The COSC triage process is the sorting of Soldiers based on an assessment of their needs and capabilities, and the location where they can best be managed in keeping with BICEPS principles. Triage is applicable at every level of care. The two key components of COSC triage are assessment and disposition.

8-2. Assessment is an evaluation of the Soldier's physical and BH needs, potential medical emergencies, and other safety risks. Assessment is performed by COSC personnel according to their professional training, expertise, and standards.

8-3. Disposition is the COSC intervention plan to address the needs identified in the assessment. Disposition has two components that include—

- Determining what intervention techniques best address the Soldier's needs and functional capabilities.
- Selecting the best location where the Soldier can be managed. The personnel conducting COSC triage should consider the needs, abilities, and the safety of the Soldier. They should also consider the unit's capacity to provide COSC interventions based on its OPTEMPO mission, resources, response to prior consultations, and willingness to participate in COSC interventions.

TRIAGE ALGORITHM FOR COMBAT AND OPERATIONAL STRESS CONTROL

8-4. Like surgical triage categories, COSR also uses triage categories. Each of the COSR triage categories are discussed in detail in the below paragraphs. The COSC triage algorithm presented in Table 8-1 uses some of the triage categories.

Table 8-1. The combat and operational stress control triage algorithm

Step 1	Is this a medical emergency?				
	Yes	Refer to nearest MTF	No	Go to Step 2	
Step 2	Does the Soldier require medical/behavioral observation?				
	Yes	Go to Step 3 A	No	Does Soldier have presumptive COSR or MH disorder?	
				Yes	Go to Step 3 B
				No	Help-in-place
Step 3	A. Can the Level II MTF or CSC Soldier Restoration Center provide adequate evaluation and intervention?		B. Can the Soldier's unit support the 5 R's or other treatment interventions?		
	Yes	HOLD*	Yes	UNIT	
	No	REFER	No	Is there a suitable support unit?	
				Yes	REST
				No	HOLD*
* When deciding between two or more potential Level II MTFs or CSC Restoration Centers, refer the Soldier to the one closest to his unit that meets his COSC needs.					
Step 4	Use on subsequent triages. Has the Soldier improved after appropriate duration of intervention?				

TRIAGE CATEGORIES FOR COMBAT AND OPERATIONAL STRESS REACTION CASES

8-5. The following are triage categories that may be used for COSR cases. *Help-in-place* (HIP), *rest*, *hold*, and *refer* cases are discussed below.

HELP-IN-PLACE CASES

8-6. *Help-in-place* is used to identify those cases that do not have severe COSR or BH disorders. They are provided COSC consultation and education, as appropriate, and remain on duty. These interactions may occur in any setting (for example, dining facility, workplace, or the post exchange). Individual identifying information is not retained or documented. There is no implicit or explicit therapist-patient or therapist-client relationship in HIP interactions.

8-7. The unit identifies those cases that remain with or return to their original unit, either for full duty with their section/platoon or for light duty with extra rest and replenishment within a headquarters element. This option depends on the unit's mission, resources, and the Soldiers symptoms. Personnel performing triage must, therefore, be familiar with the unit's situation and take that into account. When the Soldier's condition improves, the Soldier and/or unit may not feel that additional triage is necessary.

Note. Help-in-place is a new triage category that replaces the old *duty* category and is also used for unit cases.

REST CASES

8-8. *Rest* identifies those cases that are provided rest and replenishment in a nonmedical support unit, usually one that is in support of their unit. These Soldiers do not require close medical or BH observation or treatment. They are unable to return immediately to their own unit either because their unit cannot provide an adequate environment for the 5 R's; or transportation is not available for at least a day; or the 5 R's can best be coordinated from the nonmedical support unit. This option depends on the resources and mission of the available CS/CSS units, as well as on the Soldier's symptoms. Someone must be designated to be in charge of the Soldier and ensure that the 5 R's are provided. There must be a reliable transportation link to return the Soldier to his original unit after a day or two. When the Soldier's condition improves sufficiently for him to return to his unit, the Soldier and/or the supporting unit may feel that additional triage is necessary.

HOLD CASES

8-9. *Hold* refers to those cases that require close medical/BH observation and evaluation at a Level II MTF or COSC Soldier restoration center because the Soldiers symptoms are potentially too disruptive or burdensome for any available CSS unit or element. Soldier's symptoms may be caused by a BH disorder that could suddenly turn worse and require emergency treatment. The Level II MTF or COSC Soldier restoration center must have the capability to provide the necessary medical observation, diagnostic tools, and adequate stabilization for emergency treatment. When deciding among capable Level II MTFs or COSC Soldier restoration centers, refer the Soldier to the one closest to his unit that meets his COSC needs. Assessment of closeness considers speed and reliability of transportation and back. Consider transferring to another Level II MTF or COSC Soldier restoration center with increased capabilities before changing a Soldier's triage category to *refer*. All *hold* cases will be triaged again by COSC personnel or other trained medical personnel after they have been placed in this category.

REFER CASES

8-10. *Refer* cases are similar to the *hold* cases, except that *refer* cases are too disruptive and burdensome for the MTF or the COSC Soldier restoration center that is not resourced to care for this particular case. The MTF or COSC Soldier restoration center cannot provide the necessary level of diagnostic and treatment capabilities. *Refer* cases requiring care at a COSC reconditioning center, a Level III MTF or higher levels of care will be triaged by COSC or other trained medical personnel prior to being transferred to these facilities.

TRIAGE PERSONNEL FOR COMBAT AND OPERATIONAL STRESS CONTROL

8-11. All COSC personnel participate in the triage process according to their professional training, experience, and standards. Familiarization training among BH disciplines extends the effectiveness of all COSC personnel in triage skills. Medical care providers must be mentored to use the COSC triage process. A commander providing the 5 R's and placing a Soldier in a support unit for a temporary break does not equate to COSC triage.

TRIAGE CONSIDERATIONS

8-12. Triage should be initiated when the—

- Soldier is a self-referral.
- Chaplain has referred the Soldier.
- Medical personnel have requested a COSC consultation and referred the Soldier.

- Unit member/buddy has referred the Soldier.
- Leader has requested an informal referral.
- Soldier is a command-directed referral (see DODD 6490.1 and DOD Instruction [DODI] 6490.4).
- Combat operational stress control personnel observe a Soldier's behaviors which indicate possible COSR or a BH disorder.

8-13. Factors that influence an assessment may vary in depth and duration due to several other factors. These factors may include the—

- Referral source.
- Nature of the complaint.
- Observed needs.
- Medical/psychiatric history.
- Availability of resources.
- Amount of privacy for conducting assessment.
- Operational Tempo.
- Environmental conditions.
- Professional training of the person making the assessment.
- Command interest.
- Soldier cooperation.

Note. Regardless of these factors, COSC personnel are responsible for conducting the assessment in a timely manner within professional standards of practice.

DOCUMENTATION

8-14. An assessment is documented according to AR 40-66 whenever the Soldier—

- Is diagnosed with a BH disorder.
- Has a condition (or suspected condition) requiring emergency medical evaluation or treatment.
- Is prescribed medication.
- Is assessed to be a high safety risk (for example, homicidal or suicide, cognitive impairment, substance abuse, and impulsivity).
- Requests that documentation of his assessment be made in his medical records.
- Is evacuated beyond Level II MTF for further assessment or treatment.
- Is command-referred for a BH evaluation.

TRANSFER AND EVACUATION

8-15. All COSC personnel are responsible for knowing the transfer/evacuation policies and procedures within their AO. Policies and procedural information are available through the command surgeon, medical regulating officer (MRO) of the medical command, control, communications, computers, and intelligence (C4I) headquarters or major MTF. All relevant background and/or clinical documentation must accompany the Soldier during the transportation or evacuation process.

MODE OF TRANSPORTATION

8-16. Nonambulance transport is the preferred mode of transportation for COSR and nonurgent BH cases. Examples of nonambulance transportation include the Soldier's unit vehicles, supporting supply/logistics vehicles, and nonambulance medical vehicles. Ambulances convey patient status on Soldiers and often must be reserved for medical emergency. Under the provisions of the Geneva Conventions, ambulances must be used exclusively in the performance of humanitarian duties, therefore, they cannot be used to return Soldiers to duty and transport must be provided by the supported unit.

ESCORT

8-17. Depending on a Soldier's condition, an escort (either medical or nonmedical) may be necessary to provide safety, monitoring, and accountability during transportation or evacuation. The escort should be an NCO or officer of equal or greater rank/grade as that of the escorted Soldier. Escorts must be emotionally mature, responsible, and capable of conducting their escort duties. Frequently escorts carry the Soldier's clinical documentation to the destination MTF.

FEEDBACK

8-18. Good communication is essential for effective continuity of care during the transportation/evacuation process. The Soldier's unit must be informed about his location and status throughout the process. The sending party must provide sufficient documentation about the Soldier's condition, history, and administered interventions. The receiving party must provide feedback to the sending party regarding receipt of the Soldier and his documentation.

SECTION II — PRECAUTIONS AND DIFFERENTIAL DIAGNOSTIC PROBLEMS ASSOCIATED WITH COMBAT AND OPERATIONAL STRESS CONTROL TRIAGE

PRECAUTIONS FOR COMBAT AND OPERATIONAL STRESS CONTROL TRIAGE

8-19. Medical emergencies must be identified during COSC triage. Medical emergencies consist of physical illnesses or injuries and/or BH disorders that can result in permanent injury, disability, or death. Early identification of a medical emergency avoids unnecessary delay in definitive treatment. Medical emergencies can cause emotional and/or BH changes and may resemble COSR in presentation. The following conditions and behaviors could be medical emergencies. Therefore, it is important to ensure medical examinations and disposition of Soldiers that display the following conditions—

- Psychosis.
- Mania.
- Alcohol withdrawal.
- Substance intoxication.
- Delirium.
- Suicidal gesture, attempt, or high risk for suicidal behavior.
- Catatonia.
- Significant paresis, paralysis, and/or sensory loss.

8-20. Assessing for physical illnesses or injuries is a critical part of COSC triage. The COSC personnel must always consider physical illnesses or injuries that resemble COSR or BH disorders. Physical illnesses or injuries may not reach the threshold of a medical emergency, but must be recognized and appropriately treated. Assessing for physical illnesses or injuries requires an adequate review of body systems and a quick physical examination. The examination includes vital signs, examination of head, eyes, ears, nose, throat, chest, abdomen, and extremities with simple testing of reflexes and muscle strength. In field situations (Level I and II), negative or normal findings need to be documented on DD Form 1380 (US Field Medical Card). Refer to AR 40-66 for detailed information on patient accountability and management of individual health records. Any positive findings from the physical examination must be evaluated further. If the examiner has not checked various body systems, it is not reassuring to tell a Soldier that his physical or BH complaints are only COSR. All COSC personnel should receive familiarization training on basic medical examination techniques and in documenting medical information. Whenever a physical illness or injury is suspected, personnel should consult with their medical peers for further assistance. Some cases will require direct medical examination by a physician or physician assistant (PA). The COSC personnel should not order tests or procedures that do not directly influence case management. Medical tests may promote the patient role in the mind of the Soldier. Needless tests may delay a Soldier's RTD and encourage secondary gains.

DEFER DIAGNOSIS OF BEHAVIORAL DISORDERS

8-21. During assessment, COSC personnel must always consider BH disorders that resemble COSR, but defer making the diagnosis. The COSC personnel favor this default position to preserve the Soldier's expectations of normalcy (according to BICEPS). This is also done to avoid stigma associated with BH disorders and to prevent the Soldier identifying with a patient or sick role. Deferral is also preferred because some diagnoses require extensive history collection or documentation that is unavailable during deployment situations (such as personality disorders and attention deficit hyperactive disorder). It is possible that a Soldier can have a combination of COSR, BH disorders and physical illnesses/injuries at the same time. In such cases, COSC personnel must rely on their clinical experiences, training, and consultation with peers and medical personnel to distinguish among these sometimes overlapping conditions. Physical injuries/illnesses are treated at a MTF, however, the Soldier may return for further COSC interventions and activities. Deferral of diagnosis is preferred, but diagnosis can be considered if the Soldier—

- Presents for reemerging symptoms of a previously diagnosed and/or treated BH disorder.
- Presents for refill prescription of psychotropic medication.
- Has a medical condition or BH disorder listed above in paragraph 8-19.
- Is enrolled in a reconditioning program.
- Fails to improve after having received four to five days of continuous COSC interventions and activities at hold or refer status.
- Requires individual BH treatment.
- Is referred for multiple episodes of COSR.

DIAGNOSTIC CONSIDERATIONS FOR DIFFERENTIAL DIAGNOSTIC DISORDERS

Low-Grade Environmental or Stress-Related Illnesses

8-22. Low-grade environmental or stress-related DNBI illnesses can drain the Soldier's strength and confidence. For example, chronic diarrhea and slight fever may exhaust, demoralize, and contribute to COSR among Soldiers. These conditions should be treated medically, concurrently with physical replenishment, rest, reassurance and organized activities, which restore the Soldier's confidence. If they persist in spite of rest and symptomatic treatment, a more aggressive workup and treatment may be indicated.

Dehydration

8-23. Dehydration deserves special mention because it can be very insidious. Soldiers under battlefield or heavy work conditions become extremely dehydrated without feeling thirsty. This is especially likely in CBRN equipment, or in a desert/arctic environment. Refer to FM 4-25.10 for additional information on the prevention of dehydration and FM 4-25.11 for first aid measures.

Hyperthermia

8-24. Hyperthermia (overheating) in an otherwise healthy individual often first causes mild elation and excessive energy. This may be followed by irritability, disorientation, and confusion. When core body temperature climb above 106° Fahrenheit (F) or 41° Centigrade (C), the Soldier may become belligerent, combative, and have visual hallucinations. If brain temperature rises further, the Soldier collapses and convulses in heatstroke. Refer to FM 4-25.11 for first aid measures and FM 4-25.10 for additional information on prevention and first aid measures for heat injuries.

Hypothermia

8-25. Hypothermia may cause an individual to become disoriented when core body temperature falls below 95°F (34.6°C). The person may move and speak slowly. His skin looks and feels warm, leading him to take off clothing. He becomes disoriented, then unresponsive and may appear to be dead. Hypothermia is as likely in cool or even warm, wet climates as it is in extremely cold ones. Refer to FM 4-25-10 for additional information on prevention and FM 4-25.11 for first aid measures.

Overuse Syndromes

8-26. Overuse of muscles, joints, and bones that have not been prepared for the strain of field duties can lead to persisting stiffness, pain, swelling, and orthopedic injuries. If severe, these injuries may require evacuation to a hospital for evaluation. Even if these injuries are avoided, the unfit person who overexerts has days of stiffness, aching, and weakness. Such cases are likely to develop COSR if further demands are made on them.

Rhabdomyolysis

8-27. Rhabdomyolysis is one potentially dangerous complication of severe muscle overuse (and of heatstroke or crush injuries) in which myoglobin from damaged muscle cells injures the kidneys. It can cause fatigue, seizures, muscle tenderness, and muscle aches. A warning sign is dark (tea-colored) urine, but without laboratory testing, this is not easily distinguished from the concentrated urine of dehydration. Rhabdomyolysis is a medical emergency.

Head Trauma

8-28. Concussion may stun the individual and cause amnesia, residual confusion, and/or impulsive behavior. For any case of suspected head trauma or for any case of significant memory loss (especially for a discrete period of time), check scalp, eyes, ears, nose, cranial nerve signs, and vital signs for evidence of head injury. Cases with deteriorating mental status are medical emergencies. If one pupil becomes larger than the other, it is an extreme emergency requiring immediate hospitalization. Left untreated, the condition can progress rapidly to coma and respiratory arrest within hours. If a head injury is suspected, monitor mental status and vital signs periodically, especially respiration, even though physical findings are negative. Awaken the Soldier periodically to check mental status and pupil size (allowing sufficient time to recover from any grogginess on awakening). Continuous monitoring is appropriate if there are serious concerns about the risk.

Spinal Cord Trauma

8-29. Pressure, bruising, and hematomas of the spinal cord, as well as severing of the spinal cord, can cause spinal shock, with loss of sensory and/or motor functions below the level of the injury in the affected dermatome and muscle group patterns. The loss of function may be bilateral, unilateral, or partial. These cases could be confused with paralysis or sensory loss presentations of COSR. Further manipulation of a fractured spine can worsen or make permanent the spinal cord damage. Information from the history of onset, a cautious physical and neurological examination, or complete relief of symptoms following hypnosis or strong positive suggestions could demonstrate convincingly that this is only COSR. It is best to be cautious and keep the spine immobile during care and transportation.

Postconcussion Syndromes

8-30. Postconcussion syndromes may persist weeks to months beyond the period of acute concussion. Postconcussion syndromes may include perceptual or cognitive impairment, poor impulse control, and difficulty in planning ahead. These are often accompanied by cranial nerve deficits or soft neurological signs.

Abdominal Trauma

8-31. Ruptured spleen or other intraperitoneal bleeding may cause shock. The Soldier may arrive in a fetal position and be unresponsive but have reflex guarding due to peritonitis.

Air Emboli and Focal Brain Ischemia

8-32. High blast overpressures from incoming high explosive ordnance can produce air emboli (bubbles in the blood) and focal brain ischemia (small areas in the brain which cannot get oxygen because the blood flow has been interrupted). Nuclear explosions can do this, as can high explosives when shock waves are magnified by reflection within bunkers, buildings, and trenches. Ruptured eardrums, general trauma, and evidence of pulmonary damage should be detectable. Cases may have stroke symptoms (loss of muscle strength, loss of sensation in parts of the body, and/or speech disturbances), which may be subtle and mistaken for COSR.

Laser Eye Injury

8-33. Laser range finders/target designators cause small burns on the retina if they shine directly into the eye, even at great distances and especially if viewed through optics. If the laser beam causes a small retinal blood vessel to bleed inside the eyeball, the person will see red. If blood inside the eye is confirmed on examination, the Soldier should be evacuated to a hospital with verbal reassurance that he may RTD soon. If the laser does not hit a blood vessel, the Soldier may see only flashes of light, followed quickly by some painless loss of vision. If the laser damage areas of the eye responsible for peripheral vision, the Soldier may never recognize a visual deficit. If the Soldier was looking directly at the laser source however, there will likely be a major loss of visual clarity. With simple retinal burns in the retina's periphery, most of the visual symptoms recover with hours to days of rest, reassurance, and nonspecific treatment the same as with COSR. Calm, professional treatment at each level of medical care should emphasize that the injury is not vision-threatening and the chances for some, if not total, recovery is good. Soldiers with the simple retinal burns should provide self-care to decrease the risk of assuming a patient mind-set and to promote their chances of RTD. For additional information on the management of laser eye injuries, refer to FM 8-50.

Middle Ear Injuries/Diseases

8-34. Temporary loss of hearing can be cause by a decreased acoustic sensitivity following a brief extremely intense noise (explosive) or less intense, longer duration noise. Tinnitus (ringing in the ears) can also result from acoustic nerve damage or irritation, as well as from high doses of certain drugs, such as aspirin. Distinguishing physiologic from psychogenic hearing loss may require consultation with an otolaryngologist (ears, nose, and throat specialist).

Peripheral Neuropathies

8-35. Peripheral neuropathies include compression neuropathies, which are especially likely in military settings (for example, *rucksack palsy*). Depending on severity, they may require temporary job reclassification during convalescence. As these injuries are not life-threatening, a hasty diagnosis should not precede a trial of Soldier restoration treatment.

Uncommon Endemic Neurologic Disorders

8-36. These physical diseases can first manifest with cognitive emotional and/or behavioral symptoms. A comprehensive neurological examination is required for the definitive diagnoses. Examples include—

- Guillain-Barre Syndrome manifests with muscle paralysis, usually without sensory loss, which ascends the legs and arms, then the trunk, over hours to days. It is sometimes triggered by immunizations, as might be given to troops deploying overseas. It often progresses to a life-threatening situation as the muscles of respiration become involved. This requires evacuation to EAC and continental United States (CONUS). Fortunately, recovery is usually complete, but it takes months to years.

- Multiple sclerosis is a disease that can mimic many types of COSR/BH disorders with its sometimes transitory, shifting motor, sensory, speech, and cognitive/emotional symptoms. It is made worse by stress and may be difficult to diagnose. Once confirmed, multiple sclerosis cases should be evacuated to CONUS, as should other rare, progressive diseases like Lou Gehrig's disease (amyotrophic lateral sclerosis).
- True convulsive seizure can manifest after head injury or a sublethal or chronic nerve agent exposure. These are treated with normal anticonvulsant medications. Fear of nerve agent exposure may lead some Soldiers to experience psychogenic seizures. These psychogenic seizures are also called pseudo-seizures. In addition to falling unconscious and convulsing, urinary and fecal incontinence can occur during a pseudo-seizure.

SUBSTANCE ABUSE/DEPENDENCE

ALCOHOL

8-37. Substance abuse is an example of misconduct stress behaviors and not necessarily COSR. Drug and alcohol abuse may occur in active CZs and nearby areas where use is explicitly prohibited and severely punished. Personnel performing the COSC assessment should be familiar with evaluation and treatment of substances abuse and dependency. Combat operational stress control providers should consider the following:

- Heavy habitual use of alcohol, even by otherwise capable officers and NCOs, may go unnoticed in peacetime. However, alcohol abuse it may degrade necessary mission performance demanded by combat and may result in withdrawal symptoms when access to alcohol is interrupted. Alcohol withdrawal is potential medical emergency; consultation with medical personnel is essential.
- Intoxication or withdrawal from alcohol, barbiturates, and tranquilizers may be mistaken for COSR or another BH disorder. Intoxication or withdrawal requires medical treatment. Withdrawal seizures or impending or ongoing delirium tremens need emergency medical treatment.

OVERUSE OF STIMULANTS

8-38. Stimulants may cause panic attacks, hyperactivity, mania, rage attacks, psychosis or paranoia. Cessation of amphetamines after prolonged use causes a crash characterized by extreme sleepiness, lethargy, overeating depression and suicidal thinking. This condition may require one to two weeks of hospitalization to assure safe recovery.

HALLUCINOGENIC DRUGS

8-39. Hallucinogenic drugs cause sensory distortion, panic, bizarre thoughts, and potentially dangerous behaviors. These drugs may be employed by the enemy as chemical or biological warfare agents. Phencyclidine hydrochloride (PCP) is especially problematic since it also blocks pain and tends to make those under its influence paranoid, violent, and abnormally strong. Hallucinogenic drug psychosis should not be treated with antipsychotic drugs. Physically restrain and sedate patients as necessary.

INHALATION OF FUMES

8-40. Inhalation of fumes (either by accident or as deliberate abuse) and carbon monoxide poisoning can cause disoriented, abnormal behavior. Supportive treatment and, specific antidotes/medication may be needed.

ANTICHOLINERGIC DELIRIUM

8-41. In combat, atropine delirium may occur. Soldiers are equipped with atropine injectors to use as first aid against nerve agents. Two milligrams (mg) (one atropine injector) without nerve agent challenge can cause rapid pulse, dry mouth, slightly dilated pupils, decreased sweating (hot, dry, flushed skin), and urinary retention. In some individuals, 6 mg of atropine (equal to three atropine injectors) may cause hallucination and disorientation in the absence of a nerve agent challenge. Such side effects may be more pronounced in sleep-deprived Soldiers. Overheated Soldiers are more susceptible to the atropine side effects. Atropine compounds the complications of overheating by diminishing the body's ability to lose heat through sweating. One dose (2 mg) of atropine can reduce the efficiency of heat-stressed Soldiers. Two doses (4 mg) will sharply reduce combat efficiency, and 6 mg will incapacitate troops for several hours. Some plants can also cause anticholinergic delirium when eaten.

ANTICHOLINESTERASES

8-42. A nerve agent is an anticholinesterase similar to many insecticides. Low-dose nerve agent exposure may produce miosis (pinpoint pupils) without other signs. Miosis decreases vision except in very bright light and may cause eye pain when attempting to focus. This miosis may take hours to days to improve spontaneously, depending on the degree and type of exposure. Evidence gathered from farm workers poisoned by insecticides suggests that mild personality changes, insomnia, nightmares and chronic persistent depressive symptoms may be seen even after use of an antidote. Low-dose nerve agent exposure may lower the seizure threshold of many Soldiers. True epileptic seizure cases must be distinguished from those Soldiers who may have pseudo-seizures.

BEHAVIORAL DISORDERED PATIENTS IN THE THEATER

PRIMARY BEHAVIORAL DISORDERS

8-43. Primary BH disorders (especially schizophrenic-form/schizophrenic disorder, major depression, and bipolar disorder) will continue to occur at approximately the same rate as in peacetime. Some Soldiers may hide their disorders by receiving care through civilian channels. Once in the theater they may experience a relapse or self-refer to a MTF when their medication supply is exhausted.

PERSONALITY DISORDERS

8-44. Preexisting personality disorders may make a Soldier unable to adapt to military life. However, studies have failed to show a relationship between personality disorders and the likelihood of breakdown in combat. Once Soldiers with personality disorders have developed COSR or a BH disorder, they may have greater difficulty recovering and RTD. Diagnosis should never be made in haste; diagnostic criteria must be supported with adequate historical information.

Chapter 9

Combat and Operational Stress Control Stabilization

SECTION I — EMERGENCY STABILIZATION

STABILIZATION

9-1. Emergency stabilization is the acute management of disruptive behavior resulting from COSR and/or a behavioral disorder. The disruptive behavior severely impacts unit functioning by posing a danger to self and/or others. In some cases, an underlying medical condition leads to the disruptive behavior and may present an additional threat to the Soldier's life. Emergency stabilization consists of interventions that temporarily reduce a disturbed Soldier's threat of self or others, thereby allowing further medical evaluation and/or treatment. Some behavioral disorders are associated with violent behavior, such as psychotic disorders, bipolar manic disorders, antisocial personality disorder, and borderline personality disorder. Violent behavior is also associated with disruption of brain functioning due to organic factors such as intoxication, hyperthermia, metabolic imbalance, or CBRN exposure.

9-2. The COSC triage process will be repeated throughout the emergency stabilization and will determine the disposition of the Soldier. After emergency stabilization, subsequent triage could result in an immediate RTD, transfer to a COSC Soldier restoration program for observation, or further evacuation.

METHODS USED FOR EMERGENCY STABILIZATION

9-3. Methods that may be used for emergency stabilization include—

- Providing verbal reassurance and reorientation are the best methods for controlling an agitated or disruptive Soldier. If these fail, a nonthreatening show of strength or force may suffice, or sedating medications may be offered to the Soldiers. If all other means fail to reduce the threat to self and/or others, physical restraint must be considered. Given the risk for violence, it is inadvisable to attempt subdue/restraint method one-on-one.
- Applying physical restraints is reserved for subduing and restraining agitated or disruptive Soldiers who fail to respond to safer and less restrictive forms of restraint (for example, verbal warnings or show of strength). Placing a disturbed Soldier into physical restraint increases the risk of injury to the Soldier and restraint team. Prolonged or improper application of physical restraint can cause injury to the disturbed Soldier. Given the potential for injury, it is paramount that COSC personnel receive training in proper physical restraining methods. Safe medical evacuation by ground (preferred) or air ambulance.
- Regardless of the method, the restrained Soldier must be checked frequently to guard against nerve injuries or impaired circulation, which may lead to skin ulcers or gangrene. It is important to check periodically to ensure the Soldier is not secretly escaping from restraints. The Soldier is provided verbal reassurances with positive expectations for his recovery each time he is checked.
- Chemical restraints (such as medication) can be administered to a disturbed Soldier to reduce the risk of harm to self or others. Medication may be offered to the Soldier in conjunction with verbal reassurances and reorientation. Chemical restraints may only be ordered by a medical professional who is authorized to prescribe medication when a Soldier is incompetent to make medical decisions for himself and/or when the Soldier's behavior places himself or others in danger. Once administered, medical personnel must observe for side effects and adverse reactions, and must consider administering additional medications as needed. When a Soldier is in physical restraints, medication may no longer be essential, but may serve to reduce the risk of escape, limb damage, and overheating. As a secondary benefit, once the medications reduce the

Soldier's agitation, others in the vicinity may feel safer and calmer. Before prescribing an antipsychotic medication, there are a few things to consider. First, some antipsychotic drugs may take several hours or days to take effect. Second, early administration of an antipsychotic drug may confuse the clinical picture for the next evaluator in the evacuation chain. The recommendation for most cases is to use no medication unless it is truly necessary for management.

SECTION II — COMBAT AND OPERATIONAL STRESS CONTROL FULL STABILIZATION

FULL STABILIZATION

9-4. Full stabilization is normally the mission of the Level III MTF such as a CSH specialty clinic's psychiatric service. Full stabilization goes beyond securing the safety of the Soldier and those around him. It provides a safe environment for the Soldier to receive treatment interventions, continued evaluation, and assessment for RTD potential. If RTD within the evacuation policy is not feasible, the full stabilization process helps to prepare the patient for a safe, long-distance evacuation. If a Level III MTF is not available, full stabilization may be accomplished by COSC personnel when appropriately supported or by using equipment diverted from the COSC Soldier restoration capability. Considerations for full stabilization may include—

- Conducting full stabilization for NP patients is desirable for the sake of the Soldier's future treatment and for the potential of returning some Soldiers to duty. However, full stabilization is personnel intensive with a relatively low RTD payoff. Providing only sufficient stabilization to allow evacuation from the theater may be acceptable in order to maintain the other COSC functions.
- Ensuring appropriate timely evacuation of Soldiers with NP/behavioral disorders according to the theater evacuation policy. It is preferred that full stabilization is achieved for all NP patients to facilitate appropriate and timely evacuation according to the theater evacuation policy.
- Assessing and triaging of COSR Soldiers undergoing full stabilization is an ongoing process. In subsequent triages, if a Soldier becomes stable the potential for RTD, they may be transferred to a COSC Soldier restoration or reconditioning program, or may be RTD directly.

TENETS OF FULL STABILIZATION

9-5. The COSC full stabilization includes ongoing evaluation of RTD potential. This requires assessment of mental status and performance capability overtime without excessive drug effects or limitation on activity. Contact with the Soldier's unit may be needed to get information on prior history and functioning. The further from the unit the Soldier has been evacuated, the more difficult it may be to contact the Soldier's unit. Full COSC stabilization normally takes several days. To the extent compatible with safety, the stabilization program should adhere to the principles and methods for treating COSR and behavioral disorder (such as BICEPS and 5 R's).

9-6. During full stabilization, special efforts should be made to maintain and reinforce the Soldier's identity as a Soldier. Techniques that may be helpful in maintaining the Soldier mind-set include—

- Keep Soldiers in duty uniform, not pajamas, as soon as they can be allowed.
- Maintain rank distinctions and appropriate military courtesy.
- Encourage self-care and helping behaviors.
- Engage in military work activities appropriate to the Soldiers' level of function and MOS.

9-7. Initial and ongoing assessments are essential to tailor the treatment to the Soldier's individual needs. It is essential that clinical documentation is available for these assessments. The Soldier's condition is an evolving one, and must be monitored throughout full stabilization. If assessed capable to RTD, efforts should be made to return the Soldier to duty.

9-8. Ongoing treatment and/or therapeutic modalities are essential to improving a Soldier's chances to RTD whether in theater or after evacuation. Therapeutic modalities are similar to those used on inpatient units, but must remain consistent with COSC principles. These modalities include medication, individual psychotherapy, group psychotherapy, and appropriate therapeutic occupations. Observed responses to therapeutic modalities allow informed recommendation for RTD status.

FULL STABILIZATION FACILITIES

9-9. Full stabilization is commonly conducted in the CSH. The CSH can provide more sophisticated procedures, laboratory and x-ray capabilities than are available at a medical company MTF. If a medical unit, CSC is providing emergency or full stabilization, the Soldier must be kept separate from other Soldiers in Soldier restoration or reconditioning.

9-10. If the Level III MTF (CSH) cannot provide sufficient inpatient psychiatric treatment for Soldiers requiring stabilization and preparation for evacuations, the theater/AO COSC consultant may recommend up to two temporary COA until the shortfalls resolve. He can recommend to the higher medical command that COSC personnel from one or more medical unit, CSC augment the Level III MTF psychiatric service until the caseload decrease or until replacements or additional COSC personnel are brought into theater or the MTF. Lastly, the medical C2 headquarters may direct that a COSC Soldier restoration or reconditioning assets be collocated with the Level III MTF to provide an "overflow" ward as well as augmenting staff. Definitive information is provided on the CSH NP ward staff capabilities in FM 4-02.10.

9-11. The MF2K CSH has a psychiatrist, three psychiatric nurses and a medical/surgical nurse, a social worker, nine mental health specialists and an OT sergeant at full strength. It has a modular hospital tent to provide a 20-bed hospital ward. The MF2K hospitals were designed to be at one location, but may follow the MRI doctrine of splitting off a portion to a remote site, taking with it some of the psychiatry service personnel. One portion would not have the psychiatry ward module, and would need to function similarly to the MRI CSH psychiatry service.

9-12. The MRI CSH has a psychiatrist, a psychiatric nurse, and two MH specialists. The service does not have an organic psychiatric ward where NP patients are admitted. There is normally an intermediate care ward with medical ward staff designated to receive NP patients. Some of these patients seen by the NP clinic will already be stable, and are being evaluated and prepared for RTD or evacuation. Most patients who are admitted for emergency or full stabilization will be on the designated ward, although a few with serious medical complications could require the intensive care ward. If the hospital has a high census of medical and surgical patients, most of the NP patients will need to occupy beds on the intermediate care ward. The MRI hospital may be augmented by staff from a medical company/detachment, CSC to operate a NP ward, as required.

9-13. Full stabilization facilities in theater can be categorized into two types:

- Mobile facilities use general purpose (GP) large or medium tents as used in medical units, CSC and can also be available to a CSH or a tent, expandable, modular, personnel (TEMPER) tent as used in a CSH.
- Fixed facilities use buildings that were previously hospitals or buildings converted to hospitals.

9-14. Facilities used for full stabilization include tents, Deployable Medical Systems (DEPMEDS) TEMPER tents, and fixed facilities. The adaptation of these facilities have both advantages and disadvantages that include—

- The principal advantage of the (hospital) TEMPER tents, as assembled into DEPMEDS hospital, is their climate control capability. This may be a significant safety advantage for treating seriously disturbed patients in restraints with high-dose medication, which can disrupt body's ability to regulate normal body temperature. The TEMPER and standard tents both pose greater problems for security than do fixed facilities. The staff may, therefore, have to rely more than is ideal on physical restraints and medications for sedation of some cases. Blankets or screens can be used to isolate or segregate problem patients from others. Such partitions reduce mental contagion but provide little true protection.

- Standard hospital beds are on high, lightweight metal legs with wheels. For full stabilization purposes, these should be replaced with standard low, stable cots to hold strong, agitated patients in restraints. The cots also make a more "military" setting and can be used as seats for group activities.
- When feasible, it is best to have a separate "closed" (high security) and "open" (moderate/minimal) security area. The latter could be a standard GP large tent (the same as those of the minimal care wards [MCWs] located close to the TEMPER tent of the official NP ward). The specialists (MOS 68WM6 [practical nurse] and 68W [health care specialist]) of the MCW could be given on-the-job training in supervision and military group activities for the moderate/minimum security cases if the NP staff is too small.

Chapter 10

Combat and Operational Stress Control Soldier Restoration

SECTION I — SOLDIER RESTORATION

SOLDIER RESTORATION PROGRAM

10-1. Soldier restoration is normally a 24- to 72-hour (1- to 3-day) program in which Soldiers with COSR receive treatment. Soldier restoration is accomplished using the principles of BICEPS and the 5 R's as discussed in the previous chapter. The 5 R's are tailored to the needs of the Soldier. Soldier restoration is conducted as close to the Soldier's unit as possible. Soldier restoration can be conducted by medical units/elements throughout the theater with the assistance of organic and/or augmented COSC personnel. The medical company/detachment, CSC is staffed and equipped to establish Soldier restoration programs. Soldier restoration may also be accomplished at a Level III MTF (CSH). The tenets remain the same but implementation may differ in the duration of Soldier restoration and the specialized skill and knowledge of available providers. There are three subcategories of Soldier restoration (referred to as lines of Soldier restoration) based on location and available resources that are discussed in Section II of this chapter.

SOLDIER RESTORATION PROCESS

10-2. The process of Soldier restoration involves several steps that include screening, assessment, and interventions, reintegration/coordination, and movement of Soldiers. A discussion of each of the steps in the process is provided below.

SCREENING

10-3. Adequate medical screening and treatment as necessary must be done before the Soldier begins the Soldier restoration program. Minor medical conditions can be treated during routine sick call. Soldiers entering a Soldier restoration program should be only those *hold* cases that require continuous medical and/or BH evaluation and observation for 24 hours or more. The criteria for *hold* cases are discussed under COSC triage in Chapter 8. Tending to and restoring physiological status (such as sleep and hydration) is a priority. Some Soldiers who need Soldier restoration will also have a mental disorder. Treatment for the mental disorder may continue or be initiated during the Soldier restoration process.

ASSESSMENT AND INTERVENTION

10-4. Initial assessment and subsequent COSC interventions depends upon the severity of the COSR and to what extent the reactions interfere with the Soldier's ability to function. More thorough interviews are conducted only after the Soldier's physiological status has been restored.

REINTERGRATION

10-5. Coordination is required to assist Soldiers with reintegration back into their units. Soldiers are reassessed regularly. When a Soldier experiencing COSR begins to improve, the COSC interventions shift toward the reintegration of the Soldier back into his unit. The COSC personnel must work with the COSC consultant of their area or other resources to assist with this reintegration.

MOVEMENT

10-6. Movement of Soldiers from one line of Soldier restoration to another that has greater capabilities or security may be required. When movement of these Soldiers is required, vehicles used to transport them should be nonambulance vehicles, if possible, and accompanied by escorts from their unit. If nonmedical unit members are not available to perform escort duties, medical augmentation may be required.

PRINCIPLES AND PROCEDURES OF SOLDIER RESTORATION

10-7. Initial Soldier restoration begins as close to the Soldier's unit as possible, normally near an MTF where the Soldier can get a reprieve from extreme stress but at the same time can be close to his unit. Normally, Soldier restoration facilities are established in the BSA near or adjacent to the FSMC/BSMC Level II MTF. Soldier restoration is not feasible at locations that are consistently under artillery, air, or direct-fire attack, but ideally are still within the sound of the artillery or other reminders of battle. If there is potential for attack, there must be cover and defensive positions. The location should not be one from which a move is likely within 24 hours. If there is a significant possibility of a move, only those cases that can participate actively in the move with minimal supervision should be managed at this location. The specific site of the Soldier restoration facility should be adjacent to the medical company's area and be out of the immediate (close) sight of the medical triage and treatment areas

REASSURANCE

10-8. Immediate reassurance is given to the Soldier with COSR beginning with the COSC triage as discussed in Chapter 8. Tell the Soldiers that they are temporarily joining the unit, not as patients, but as Soldiers who need a few days to recover from COSR. Emphasize that COSR is a normal response to extremely abnormal conditions. Rapid recovery is also normal and RTD is expected. Reassure the Soldiers about safety and what to do in the event of an attack or march order. Lastly, orient the Soldiers to the Soldier restoration program.

STRUCTURED MILITARY ENVIRONMENT

10-9. An emphasis should be placed on maintaining a highly structured military unit environment and schedule of activities in order to keep the Soldier from adopting a patient role. Assign the Soldier to a squad under supervision of a specific squad leader. The squad leader may be a CSC unit/MH section member or a member of the BSB subordinate unit. The squad leader may also be one of the Soldiers there for Soldier restoration, if their condition allows. The newly arrived Soldier is assigned to a tent that has been designated/erected for the Soldier restoration program. Soldiers with more dramatic COSR or BH symptoms should be temporarily quartered separately from other Soldiers receiving Soldier restoration. In remaining consistent with the principle of treating Soldiers with COSR as Soldiers instead of patients, it is recommended that Soldier restoration facilities do not display the Red Cross as displayed on the MTF. This could affect their protection under the Geneva Conventions. See FM 4-02 for information on the Geneva Conventions.

SUPPORT THE SOLDIER'S MILITARY IDENTITY

10-10. Maintain appropriate rank distinctions, titles, and military courtesies within the confines of the tactical situation. Expect the Soldier to maintain military bearing and personal appearance. They should be in duty uniform. Conduct basic Soldiering skills. Do not take personal possessions away from the Soldier. This includes weapons, unless there is significant concern for the Soldier's safety. If the Soldier arrives with a weapon, SOPs of the AO will guide whether or not the weapon is secure for the Soldier. Also, encourage the unit to maintain contact with its Soldier.

REPLENISHMENT OF PHYSIOLOGICAL STATUS

10-11. Get the Soldier under shelter and cool down if overheated, warm up if cold, and dry off if wet. Providing hot beverages and/or soup will also assist with restoring body temperature. Replenish hydration with palatable beverages and meals. Unless the Soldier is totally exhausted, institute some personal hygiene. If the Soldier needs uniforms or equipment, coordinate with the BSB S4 or supporting logistic element. Restorative sleep should be as normal as possible. Soldiers are typically able to fall asleep when reassured of safety, and do not routinely require medication to induce sleep. They should be informed that they may awaken with vivid and frightening dreams, and be instructed on quick relaxation techniques to go back to sleep. If those measures are insufficient, one-time medication with a sedative/hypnotic may be considered, coordinated, and prescribed by a physician or PA from the supporting medical company. The duration of sleep should be sufficient to make substantial progress in repaying the sleep debt. It should also begin the process of restoring a reasonable sleep/wake cycle that is consistent with the Soldier's duties in his unit.

RESTORE CONFIDENCE

10-12. Restoring a Soldier's confidence may include—

- Providing regularly scheduled formations to keep Soldiers informed of daily activities and the tactical situation, including information about their unit, when available.
- Providing therapeutic occupations that are based on the Soldier's current functional ability. The OT (or COSC team members under the guidance of the OT) selects therapeutic occupations that support the Soldier's military identity and enhance the Soldier's sense of competence. Therapeutic occupations may include activities of daily living (such as physical fitness or uniform maintenance); educational activities (selected common and collective Soldier task training or life skills training); work/productive activities (militarily relevant tasks such as vehicle maintenance or site maintenance); leisure/recreational activities (enjoyable, relaxing activities, games, and sports); social participatory activities (cooperative/competitive sports, games, ceremonies, or celebrations). Therapeutic occupations provide the casualty with a challenge but afford successful performance that shows the Soldier that he is still capable and competent. This realization plays the dominant role in restoring the Soldier's sense of confidence, functional capacity, and ability to RTD.
- Training and teaching Soldier on methods for managing excessive stress.
- Facilitating factional review of precipitating event that has caused the Soldier's COSR through talking (ventilating) and coaching. Combat and operational stress control personnel (or health care providers trained by COSC personnel) facilitate these discussions. They help Soldier restore perspective with questions and coaching. The factual review (debriefing) is often best done individually unless several Soldiers experienced the same event.

SECTION II — LINES OF SOLDIER RESTORATION

FIRST-LINE SOLDIER RESTORATION

10-13. First-line Soldier restoration is usually provided at the BSMC/FSMC in the BSA. It is provided by personnel organic to the BSMC/FSMC usually assisted and supervised by COSC personnel organic to the medical company and/or from CSC unit personnel. This also applies to a cavalry regiment. Because of the high mobility of the BSMC/FSMC, Soldier restoration will often be a 24- to 72-hour process. First-line Soldier restoration also may occur in relative secure locations in division, corps, and theater AOs. First-line Soldier restoration in the division is provided near and supported by a Level II MTF such as a MSMC with its organic MH and/or minimal augmentation from a CSC unit. First-line Soldier restoration in corps and theater is provided by ASMCs with responsibility for their AO. It is provided by COSC personnel organic to the medical company and/or from CSC unit personnel. The MSMC and ASMCs will be relatively unlikely to move on short notice so Soldier restoration up to 72 hours should be feasible. Support provided for Soldier restoration is a responsibility of the MH sections of supporting medical companies. Tasking for support of Soldier restoration programs is accomplished through the command surgeon.

SECOND-LINE SOLDIER RESTORATION (SOLDIER RESTORATION CENTER)

10-14. Each of the first-line Soldier restoration locations listed above should be backed up by a second-line Soldier restoration capability at a location that is less likely to have to move on short notice and has more COSC capabilities. Soldier restoration centers may be located near the MSMC, ASMC, or with a CSC medical company or detachment. They can be located in the BSA near the BSMC/FSMC when it is at a stable base camp. The Soldier restoration center may receive *hold* cases that are transferred from the first-line medical companies, as well as Soldiers from nearby units. Second-line Soldier restoration has more equipment and a greater range of COSC expertise. This permits a 72-hour holding capacity for a stable well-organized Soldier restoration center and may provide full stabilization.

THIRD-LINE SOLDIER RESTORATION

10-15. In some scenarios, units with Soldiers in need of Soldier restoration may be significantly closer to a Level III MTF such as a CSH than to a medical company or CSC unit. On order, a Soldier restoration program may be conducted by the CSH specialty clinic NP staff, which may also be augmented with personnel from a medical company/medical detachment, CSC. In such cases, a Soldier restoration program may be conducted by the CSH specialty clinic NP staff. It may be augmented with personnel from the medical company or detachment, CSC. Soldier restoration near a CSH must be kept clearly separate from the patient ward. It may be done at a MCW or in separate tents.

RETURN TO DUTY OF RECOVERED COMBAT AND OPERATIONAL STRESS REACTION SOLDIERS

10-16. Most COSR symptoms do not necessarily improve completely while the prospect of combat continues. The Soldiers should be given the positive expectation that they will RTD. Every possible effort should be made to return Soldiers to their original unit. The RTD of Soldiers following Soldier restoration depends on how near the Soldier's unit is, the availably of a means of transportation, and the tactical situation. Ideally, their units are contacted to send someone to get the Soldier or he may be returned to his unit by way of the personnel replacement company. Mental health personnel coordinate directly with the unit to which the Soldier is returning and/or with COSC, other medical personnel, or UMT supporting the unit's AO. These contacts can consult with the leaders of the Soldier's unit and facilitate the Soldier's acceptance back into his unit.

DOCUMENTATION

10-17. Soldiers receiving Soldier restoration interventions must be tracked from the initial contact until they are returned to their unit. A record must be maintained of interventions and activities provided during Soldier restoration, as well as the Soldier's response. A notation of the dates, any pertinent medical data, and providing unit is entered in the Soldier's medical record. An administrative summary of Soldier restoration services may be developed, stored, and disposed of as directed by the Soldier restoration center SOP. If the Soldier received BH treatment, documentation should be maintained as delineated in AR 40-66. A statistical record is maintained and sent to higher medical headquarters according to SOP.

Chapter 11
Behavioral Health Treatment

SECTION I — BEHAVIORAL HEALTH CARE

TREATMENT FOR BEHAVIORAL DISORDERS

11-1. Behavioral health treatment exists when there is an explicit therapist-patient or therapist-client relationship. Behavioral health treatment is provided for Soldiers with behavioral disorders to sustain them on duty or to stabilize them for referral/transfer. This is usually brief, time-limited treatment as dictated by the operational situation. Behavioral health treatment includes counseling, psychotherapy, behavior therapy, occupational therapy, and medication therapy. Treatment assumes an ongoing process of evaluation, and may include assessment modalities such as psychometric testing, neuropsychological testing, laboratory and radiological examination, and COSC providers' discipline-specific evaluations.

COMBAT AND OPERATIONAL STRESS CONTROL BEHAVIORAL HEALTH PROVIDER

11-2. There are five professional disciplines and two enlisted specialties that serve as BH/COSC providers, see Chapter 3. Army BH/COSC providers are officers who hold active individual clinical privileges granted by an MTF according to AR 40-68. Providers also include enlisted medical specialist who do not hold individual clinical privileges, but who work under direct supervision of a professional COSC provider. Behavioral health treatment is provided within the scope of the providers' clinical privilege, training, and experience.

SECTION II — BEHAVIORAL HEALTH TREATMENT PROTOCOLS AND MEDICATIONS

BEHAVIORAL HEALTH TREATMENT FOR SOLDIERS

11-3. Behavioral health treatment is provided to Soldiers with diagnosed behavioral disorders (see Chapter 8), and who require more intentions for their diagnoses. It is both inappropriate and detrimental to treat Soldiers with COSR as if they are a BDP. A therapeutic relationship may promote dependency and foster the "patient" role. Likewise, medication therapy and the highly structured treatment modalities imply the "patient" role. Medication for transient symptom relief (insomnia or extreme anxiety) may not be detrimental if there is no expectation that medication will continue to be prescribed.

BEHAVIORAL HEALTH PROTOCOLS AND PROCEDURES FOR TREATMENT

11-4. Behavioral health treatment is provided within a context of preventing symptoms of behavioral disorder from progressing in severity and improving the level of functioning. The principles of BICEPS apply. Treatment should not impede activities for other prevention levels (for example, universal, selective, and indicated). Treatment should not interfere with the Soldier's duties or the unit's mission. The COSC provider should emphasize Soldier's identity as a Soldier despite having a diagnosis. Transportation of the Soldier to the provider (or vice versa) and ability to assure reliable meeting times can be limiting factors, depending on the operational environment. Medication refills must be coordinated for availability.

Standards of Treatment

11-5. Treatment standards are the same in the deployed environment as in garrison. When operational requirements dictate that clinical standards of treatment/care are waived or relaxed, it must be approved by the AO COSC consultant. (AR 40-68 provides additional information.) Treatment should be tailored to the anticipated availability of the Soldier and COSC provider. Short-term interventions are more practical than long-term commitments. If longer-term treatment is necessary, design the intervention in time-limited modules. Under no circumstances should treatment diminish the Soldier's ability to provide self-care and to defend himself. Exceptions include emergency stabilization and preparation for evacuation. In addition, the VHA/DOD Clinical Practice Guidelines website (http://www.oqp.med.va.gov/cpg/cpg.htm) offer clinicians evidence-based assessment and treatment algorithms for acute stress disorder, PTSD, and many other behavioral disorders.

Army Regulations Governing Evaluations

11-6. Fitness for duty evaluations are conducted as necessary within the priorities of the supported commanders according to AR 40-501; psychiatrists should not initiate a medical evaluation board without first ensuring the Soldier has received adequate treatment. This treatment may not be available in theater. Command-directed evaluations are conducted as necessary within the priorities of the supported commanders according to DODD 6490.1 and MEDCOM Regulation 40-38. Clinical documentation should be safeguarded according to AR 40-66 and local command policy. Treatment should be conducted in a location that is as private as possible. Information can be released to a third party if the Soldier consents. Combat and operational stress control providers need to notify command when the Soldier's safety is in question (suicidal, homicidal) or if the Soldier is removed from his unit for medical observation. Additional release of information to command is on a need-to-know basis. For any questions on release of information on COSR and NP patients, consultation with supporting JAG office is advised. Also, maintaining the Soldiers health record with clinical data is required and accomplished according to AR 40-66 and supplemental theater policy as appropriate. Treatment should occur throughout the evacuation process and follow-up is expected at home station.

Chapter 12
Reconditioning

SECTION I — RECONDITIONING PROGRAM

LOCATION

12-1. Reconditioning programs are intensive efforts to restore those Soldiers triaged as a *refer* case, but who still have good potential for RTD. Referral to reconditioning can be from Level II MTF, COSC Soldier restoration program, or from Level III MTF. Reconditioning includes the rehabilitation of Soldiers with mental disorders, such as substance abuse/dependency. Reconditioning programs are conducted by COSC personnel and Soldiers usually participate for up to 7 days. Soldier participation may be extended by a case-by-case exception to theater evacuation policy. Reconditioning is conducted only in a theater where there are adequate COSC elements and supplies. When COSC resources are needed for other functional areas, any ongoing reconditioning program reduces its scope of services or closes. Reconditioning is provided at first-, second-, or third-line centers determined by location and available resources.

PROVIDERS

12-2. Conducting reconditioning programs is a mission of the medical company, CSC and the medical detachment, CSC (MRI). Reconditioning can also be done on a small-scale by a CSC medical detachment (MF2K) under some circumstances.

SECTION II — TENETS AND PROCEDURES OF RECONDITIONING

RECONDITIONING

12-3. Reconditioning may be considered an extension of Soldier restoration. Reconditioning is similar to Soldier restoration, but with potentially longer stay, treatment strategies focus on preventing atrophy of skills and assisting Soldiers in regaining skills needed for duty. Reconditioning makes more use of BH treatment modalities.

STRUCTURED MILITARY ENVIRONMENT TO SUSTAIN SOLDIER'S IDENTITY AS A SOLDIER

12-4. Like Soldier restoration, reconditioning emphasizes a highly structured military unit environment and schedule of activities in order to keep the Soldier from adopting a "patient" role. Maintaining a military environment is even more critical when colocated with hospitals or other service support elements.

REPLENISHMENT OF PHYSIOLOGIC STATUS AND CONFIDENCE

12-5. Reconditioning initially emphasizes physical replenishment and hygiene, but later shifts the emphasis to more closely match the conditions that the Soldier should expect when RTD. In order to restore confidence in the Soldier, unit formations are held on a regularly scheduled basis, and include the Soldiers receiving reconditioning.

OCCUPATIONAL THERAPY

12-6. Therapeutic occupations may include—

- Activities of daily living (for example, hygiene, physical fitness, and uniform maintenance).

- Educational activities (for example, common and collective Soldier task training, and life skills training).
- Militarily relevant work details according to the current functional level and MOS/duties of the Soldier.
- Leisure/recreational activities (for example, enjoyable, relaxing activities, including cooperative physical and mental activities, and basic relaxation techniques).
- Social participatory activities (for example, games and ceremonies).
- Group training in relaxation techniques.
- Ventilation and coaching where COSC (or medics trained by COSC personnel) encourage discussion about stressors and their impact on mental well-being. This technique helps the Soldier to restore personal perspective with questions and coaching. Individual counseling and therapy may improve the Soldier's functions as well.

12-7. When reclassification recommendations are considered, an adjacent CS or CSS unit may be able to provide a job that will match the Soldier's abilities. This provides an opportunity to demonstrate the Soldier's abilities and build confidence. The reconditioning personnel may recommend to S1/G1 that the Soldier be reclassified to another MOS.

FIRST-LINE RECONDITIONING PROGRAM

12-8. First-line reconditioning programs in the corps are staffed by task-organized CSC elements from the CSC company or by an MRI CSC detachment. Reconditioning can be conducted by an MF2K CSC detachment, but would preempt its Soldier restoration capability. If the inpatient NP workload is light, a small reconditioning center may be staffed by personnel from the NP ward and consultation service of an MF2K CSH. The first-line reconditioning center usually collocates with a CSH but must maintain its separate, nonhospital identity. It should not be situated among the hospital wards, or near the morgue, triage area, or helicopter pad. Within a theater/AO, the preferred option is to have one or two reconditioning centers placed such as to allow easy access from Soldier restoration centers. Under major, prolonged combat conditions, it may be more appropriate to have one reconditioning center behind each division. The reconditioning center is dependent on the hospital for support. The reconditioning center works with the supporting hospital by sending work parties of Soldiers in the program to assist in food preparation and delivery and cleanup chores. These Soldiers may also be used for assisting with work details throughout the hospital, but must be under direct supervision of either hospital or BH personnel. The reconditioning center uses the supporting hospital's medical records section to maintain the permanent case records. Cases in the reconditioning center are counted as patients in the reconditioning center on the daily hospital census. The cases are not counted as occupied beds when reporting the hospital bed occupancy. Upon disposition from the reconditioning center, whether for RTD, retraining for other duty, or evacuation, the reconditioning center prepares the chart for further evacuation or writes the discharge summary and closes the hospital's chart.

12-9. Some reconditioning cases will be able to return to far forward CS or CSS duty. However, many of the Soldiers who need reconditioning will be unable to return to their original unit, due to combat operations or the nature of their symptoms. Soldiers who undergo a 4- to 7-day Soldier restoration program in the corps should not be crossed off the division's personnel rolls (as specified in AR 40-216). Continuing contact with the unit increases RTD rates. For Soldiers who recover but are no longer on the division's rolls, every feasible effort should be made to return them to their original units. The chief, division MH and the CSC unit teams that are attached to reinforce the division should maintain frequent contact with the reconditioning centers that support the division. Coordination efforts, through the division support command (DISCOM) sustainment brigade and the personnel replacement system by division BH personnel, should attempt to facilitate return of recovered Soldiers to their original units. If recovered Soldiers cannot be returned to their previous small unit, consider forming them into cohesive pairs or small groups which can be reassigned to a new unit together. The standard corps evacuation policy is seven days, but some Soldiers with good potential for RTD may need a few more days at the reconditioning center. The corps evacuation policy is at the discretion of the theater commander. The principle purpose of a short evacuation policy is to keep hospital beds available for mass casualties and to minimize the expense and labor required to treat serious surgical and medical cases.

SECTION III — EVACUATION POLICY

EXTENDING THE THEATER EVACUATION POLICY

12-10. If need assessment indicates that a longer period of time (more than seven days) will achieve a Soldiers' RTD, the theater COSC consultant can request permission to extend the theater evacuation policy (up to 14 days). He submits the request for authority via the commander, MEDBDE, through the corps surgeon to the corps commander. The following facts apply: The reconditioning program, because of its austerity, is not a significant logistical burden to the corps. Soldiers in reconditioning perform useful work details and perimeter defense. Lastly, the increase in RTD is important in operations such as stability and reconstruction operations, where reducing personnel attrition becomes an important factor.

EVACUATION

12-11. Reconditioning cases that do not recover sufficiently to return to some duty within the designated evacuation period are evacuated from corps to EAC. They are best transported in cargo trucks and buses, bus ambulances, or an ambulance train. In the latter two cases, they should be assigned helper tasks. Use air evacuation only if there is no other alternative.

SECOND-LINE RECONDITIONING

12-12. Second-line reconditioning is conducted in the intermediate staging base outside the CZ in the theater. This center could be at a CSH, a fixed MTF, or a CSC unit element. The second line reconditioning center continues to emphasize physical fitness, Soldier skills, work details, and individual/group counseling/psychotherapy. Cases will be retrained for CSS duties at corps and theater levels. As soon as the recovering Soldiers are ready, the retraining site can shift to on-the-job training at a nearby CSS unit.

THIRD-LINE RECONDITIONING

12-13. Third-line reconditioning is for Soldiers with COSR and/or mental disorders who did not improve sufficiently at lower lines of reconditioning but still have RTD potential. These Soldiers are transferred to a reconditioning program at a regional MEDCEN or home station MEDDAC. These reconditioning centers, like the others, must maintain a military atmosphere and provide opportunities to engage in occupational therapy. Some Soldiers may also require retraining for other duty/MOS reclassification.

Appendix A

Combat and Operational Stress Control and Religious Support

ROLE OF UNIT MINISTRY TEAM

A-1. This appendix addresses the general role of the UMT in the commander's program for COSC. The UMT, imbedded within units down to battalion level, provide immediate support to leaders in fulfilling their COSC responsibilities. The UMTs also assist in training leaders to recognize combat and operational stress identification and intervention responsibilities. In cooperation with unit medical personnel, UMTs serve as a primary referral agency to BH resources.

RELIGIOUS SUPPORT FOR COMBAT AND OPERATIONAL STRESS CONTROL

A-2. Soldiers' inner resources are generally rooted in their religious and spiritual values. In combat, Soldiers often show more interest in their religious beliefs. When religious and spiritual values are challenged by the chaos of combat, Soldiers may lose connection with the inner resources that have sustained them. The UMT is the primary resource available to Soldiers experiencing such dilemmas and is a valuable resource in assisting them as they seek to refocus their spiritual values.

Unit Ministry Team Support for Combat and Operational Stress Control

A-3. The UMT provides preventive, immediate, and restorative spiritual, emotional support and care to Soldiers experiencing COSR.

Preventive Religious Support for Combat and Operational Stress Reaction

A-4. The UMT assists in preventing COSR and misconduct stress disorders through spiritual fitness training. Ministry of presence with Soldiers, assigned Department of Army civilians, and contractors is critical. The UMT provides a stabilizing influence on personnel and assists them in strengthening and regaining personally held spiritual values. Preventative activities include—

- Worship opportunities.
- Private and group prayer opportunities.
- Religious literature and materials.
- Scripture readings.
- Sacraments and ordinances.
- Assistance to personnel and families prior to deployment, emphasizing family strengths.

Immediate Religious Support for Combat and Operational Stress Reaction

A-5. The UMT assists commanders in the identification of personnel experiencing negative reactions to combat and operation stress, COSR, and misconduct stress behaviors. The UMT works closely with the unit's leaders and medical personnel to care for COSR cases through religious support and comfort. Immediate religious support activities may include—

- Conversation focused upon fears, hopes, and other feelings.
- Prayer for fallen comrades and memorial ceremonies and services.
- Rites, sacraments, and ordinances, as appropriate.
- Sacred scripture.

Restorative Religious Support for Combat and Operational Stress Control

A-6. Following an operation, a unit may require reconstitution. Surviving Soldiers may need to rebuild emotional, psychological, and spiritual strength. Depending upon the spiritual, emotional, and physical condition of the unit's Soldiers, the organic UMT may need augmentation from higher echelons or other units. Restorative religious support activities may include—

- Worship, sacraments, rites, and ordinances.
- Memorial ceremonies and services.
- Religious literature and materials.
- Grief facilitation and counseling.
- Reinforcement of the Soldiers' faith and hope.
- Opportunities for Soldiers to talk about combat experiences and to integrate those experiences into their lives.

Appendix B

Medical Detachment, Combat and Operational Stress Control

CONCEPT OVERVIEW

BACKGROUND

B-1. The current medical units, CSC in the Army inventory were initially developed over ten years ago. The MF2K redesign resulted in the development of the MRI. Recent events in the Global War on Terrorism in both Afghanistan and Iraq have indicated a need to update these units from MRI medical detachment, CSC to a multifunctional COSC detachment. The 2003 US Army Surgeon General's Mental Health Advisory Team (MHAT) report and observations indicate a need to update the unit in terms of modularity, multifunctionality, and its working relationship with support units and command surgeons from brigade to theater levels.

PROPOSED MEDICAL DETACHMENT, COMBAT AND OPERATIONAL STRESS CONTROL

B-2. The proposed medical detachment, COSC (TOE 08660G000) consists of a detachment headquarters, a main support section, and a forward support section. The main support section consists of its headquarters and an 18-Soldier BH team made up of social workers, clinical psychologists, psychiatrists, occupational therapists, psychiatric nurses, MH specialists, and OT specialists. The forward support section consists of an 18-Soldier BH team only. Each BH team is capable of breaking into six 3-person subteams, for battalion/company prevention and fitness support activities. This provides for a total of 12 subteams for each detachment, giving supported commanders more teams and more flexibility in the utilization of those teams, yet maintaining all of the capabilities of COSC.

Operational Environment

B-3. Soldiers will continue to experience the physical and mental impacts of high stress in both combat and stability and reconstruction operational environments. Advances in technology which will increase effectiveness, and the impact of the individual Soldier, will also increase dispersion and add to isolation and stress levels of Soldiers. The enemy of the future will look for methods that will have the greatest psychological impact on our future Soldiers; future enemies will not be our technological equals and will increasingly rely on terror and nontraditional methods to unnerve, injure, and demoralize both Soldiers and civilians. There is an identified need to update COSC units in terms of flexibility, modularity, multifunctionality, and its relationship to supported units and the surgeons responsible to commanders for the BH mission. The current configuration of the COSC units resulted in an uneven use of available COSC resources and that there was a greater need for preventive and limited fitness services forward. Commanders have communicated there is a greater need for active outreach and consultation.

Assumptions

B-4. Power projection will likely remain the fundamental strategic and operational imperative of our forces for the foreseeable future. The new medical detachment, COSC will be well suited to the future force, designed to move with and support the division and corps of the future force, and division and corps equivalents. The medical detachment, COSC will be assigned to a MEDBDE, MMB, or MEDCOM and will be operationally attached to supported units.

B-5. Each BCT has a MH section with one BH officer and one MH specialist and that all division and brigade MH sections are filled. The medical detachments, COSC will be able to augment BCTs MH sections. The medical detachment, COSC will require additional logistical, finance, maintenance, personnel, legal, FHP, and administrative services support.

Implication for Combat and Operational Stress Control

B-6. Future forces must be emotionally and mentally fit before deploying and resilient in battle in order to endure the fast pace and intense lethality of the battlefield. Commanders will need to be able to apply COSC capabilities at every level quickly and effectively. Reorganization of the medical detachment, CSC to a combined, multifunctional medical detachment, COSC will provide greater flexibility and greatly increase outreach and preventive services while maintaining capability for the entire spectrum of COSC.

Operational Concept, Mission

B-7. The medical detachment, COSC may be attached to a theater or division medical C2 headquarters in order to provide COSC casualty prevention, treatment, and management on an area basis.

B-8. The AUTL (see FM 7-15) linkage includes—

- ART 6.5. Provide Force Health Protection in a Global Environment.
- ART 6.5.1. Provide Combat Casualty Care.
- ART 6.5.1.5. Provide Mental Health/Neuropsychiatric treatment.
- ART 6.5.4. Provide Casualty Prevention.
- ART 6.5.4.5. Provide Combat Operational Stress Control Prevention.

Method of Employment

B-9. The medical detachment, COSC deploys with a division-sized force into a theater or with a task force of up to 25,000 Soldiers in stability and reconstruction operations. The medical detachment, COSC is attached for OPCON to the division/corps/theater medical C2 headquarters. The medical detachment, COSC is assigned to the MMB of units that have reorganized under the future force design. It is normally tasked by the command surgeon to support units within the command. The detachment commander as authorized by his medical C2 headquarters will coordinate operations with the command surgeon or the division/corps/theater psychiatrist. In the absence of a psychiatrist, the detachment commander will advise and consult with the command surgeon and the commander as required. Subteams may be task-organized and further attached for operational control to maneuver BCTs, and will operate in and forward from BSA, working closely with medical assets and combat religious support team (CRST). The detachment commander will coordinate efforts with the division and the CRST to provide in-theater prevention programs as required (suicide prevention, reunion, COSC training, TEM, and PTE stress management).

Detachment Headquarters

B-10. Plans, supervises and monitors the activities of the medical detachment, COSC in support of the COSC mission. See Table B-1 for a listing of personnel assigned to the detachment headquarters. The commander will establish a command post (CP) collocated with the supported unit surgeon (either division or sustainment brigade) and coordinate all activities of the detachment through the surgeon section. The detachment commander provides C2 and works closely with supported unit surgeon to provide logistics and other required services for the detachment. The field medical assistant and/or the detachment sergeant also establishes and coordinates logistics and required services support through the supported command staff for all detachment personnel and operations.

Table B-1. Detachment headquarters personnel

Detachment Commander (LTC/05, AOC 05A00)
Medical Operations Officer (CPT/03, AOC 70B67, MS)
Chaplain (CPT/03, AOC 56A00, CH)
Detachment Noncommissioned Officer in Charge (NCOIC) (SFC/E7, MOS 68X40)
Supply Sergeant (SGT/E5, MOS 92Y20)
Human Resources Specialist (SPC/E4, MOS 42A10)
Wheeled Vehicle Mechanic (SPC/E4, MOS 63B10)
Patient Administration Specialist (SPC/E4, MOS 68G10)
Religious Support Specialist (SPC/E4, MOS 56M10)
Cook (PFC/E3, MOS 92G10)
Note. The position of detachment commander may be filled by a (LTC, AOCs 73A00, 73B00, 60W00, 65A00 or 66C00). For description of duties, see Section I of the TOE and Department of Army (DA) Pamphlet 611-21.

Main Support Section

B-11. Included in the main support section is the detachment headquarters personnel identified in Table B-1 above and an 18-personnel BH team. See Table B-2 for list of personnel assigned to the BH team. This BH team can break into 6 three-person subteams, each with one officer, one NCO, and one specialist to conduct prevention and limited fitness operations in the battalion/company level in a maneuver BSA. Teams are collocated with a BCT medical unit. Modularized subteams may be combined and task-organized to provide or support any other COSC function, which includes larger scale Soldier restoration reconstitution, or any other COSC or BH mission as determined by its headquarters. Tasks for the main support section BH team includes:

- Consultation and coordination with commanders at team level (battalion and company).
- Coordination with CRST.
- Prevention and fitness support activities.
- Reconstitute any other COSC function as required.

B-12. When attached for support to any medical company Level II MTF, the grouped subteams are dependent on the supporting medical company for logistical support and quarters. The subteam can provide 3-day restoration for up to 50 Soldiers. With additional logistical support and quarters from supported command, the team can accommodate additional Soldiers under surge conditions.

Table B-2. Main section behavioral health team

Team Leader/Psychiatrist (MAJ/04, AOC 60W00)
Occupational Therapist (MAJ/04, AOC 65A00)
Social Work Officer (MAJ/04, AOC 73A67)
Behavioral Science Officer (CPT/03, AOC 67D00)
Clinical Psychologist (CPT/03, 73B67)
Psychiatric Nurse (CPT/03, 66C7T)
Occupational Therapy NCO (Four)—(Two SSGs/E6 and Two SGTs/E5, MOS 68W30/68W20)
Mental Health NCO (Four)—(One SSG/E6 and Three SGTs/E5, MOS 68X30/68X20)
Mental Health Specialist (Six)—(SPC/E4, MOS 68X10)
Note. For description of duties, see Section I of the TOE and DA Pamphlet 611-21.

Forward Support Section

B-13. This section consists of an 18-Soldier BH team. It is comprised of a section leader and five other BH officers, six MH/OT NCOs, and six MH specialists. This BH team can break into six three-person subteams, each with one officer, one NCO, and one specialist to conduct prevention and limited fitness operations in the BCT medical company in the BSA. Modularized subteams may be combined and task-organized to provide or support any other COSC function, which includes larger scale Soldier reconstitution, or any other COSC or BH mission as determined by its headquarters. The forward support section is dependent on the supported unit for C2 and logistical support. See Table B-3 for list of personnel assigned to the forward support section BH team.

Table B-3. Forward support section behavioral health team

Team Leader/Psychiatrist (MAJ/04, AOC 60W00)
Psychiatric Nurse (MAJ/04, 66C7T)
Clinical Psychologist (MAJ/04, 73B67)
Occupational Therapist (CPT/03, AOC 65A00)
Social Work Officer (CPT/03, AOC 73A67)
Behavioral Science Officer (CPT/03, AOC 67D00)
Occupational Therapy NCO (Two)—(One SSG/E6 and One SGT/E5, MOS 68W30/68W20)
Mental Health NCO (Four)—(One SSG/E6 and Three SGTs/E5S, MOS 68X30/68X20)
Mental Health Specialist (Six)—(SPC/E4, MOS 68X10)
Note. For description of duties, see Section I of the TOE and DA Pamphlet 611-21.

Organizational Concept

B-14. Required capabilities require the unit to—

- Deploy to power projection platform or point of debarkation and deploy with supported unit or deploy to theater and join supported unit.
- Provide all functions of COSC from company to theater level in close coordination with and under the direction of the supported unit surgeon.
- Provide prevention, assessment, treatment, and referral BH activities.
- Provide consultation with commanders from company to theater level.
- Conduct and coordinate administrative and logistical support to sustain operations.

Basis of Allocation Total Army Analysis

B-15. The basis of allocation is 0.333 per BCT; 1 per division; 2 per theater. In support of a theater, a COSC detachment provides support on an area basis and provides additional support to the division/corps on order. See Figure B-1 for breakdown of separate standard requirement codes (SRCs) of the medical detachment, COSC sections.

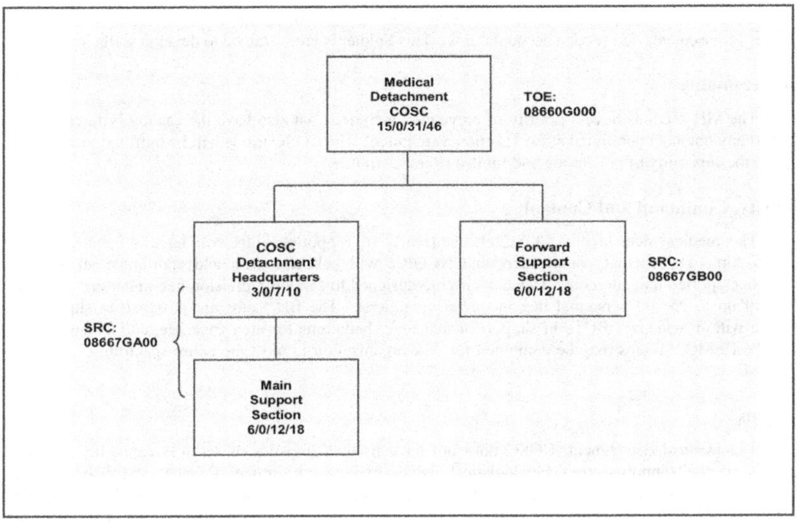

Figure B-1. Medical detachment, combat and operational stress control

Concept of Transition

B-16. Transition begins with the four active component detachments, one at a time. Personnel, transportation, and equipment changes will be minimal; transition should take a short amount of time and money to accomplish. Changes in the detachments are congruent with Force Modernization designs and personnel proponency issues. The proposed COSC detachment is designed with the Future Force in mind.

Patient Medical Records

B-17. The Surgeon General's MHAT report notes that there is a need to standardize BH reporting and documentation. The BH charts were inconsistently maintained and documentation did not always accompany patients through the evacuation chain. An additional finding was that there was no standardized method of collecting BH workload or clinical data and that no database that tracked evacuees provided reliable clinical information. The patient administration specialist (MOS 68G10) will supervise or perform functions as outlined in DA Pamphlet 611-21 for the Soldier caseload; maintain Soldier/patient accountability and records; interface with patient administration division of supported/collocated medical units.

Role of Religious Support Specialist

B-18. Findings of the MHAT include the need for chaplains to be aware of their role in COSC and that an aggressive chaplain outreach program should be executed. This requires that supervisory chaplains be involved in CRST integration with BH and primary care providers, and educate primary players in role of CRST in COSC. These activities require that the COSC chaplain be mobile and have an active presence in the maneuver brigades. This position is required to fulfill the full-time force protection requirements for the chaplain while in theater. The religious support specialist assists the COSC chaplain by screening individuals seeking counseling, coordinates for convoy security when the CRST goes out on site visits; assists chaplain with fund request; develops and coordinates needed religious support projects within the unit; coordinates with detachment for all vehicle services, communication system repairs, detachment training requirements and section accountability. This Soldier is cross-trained to develop skills for COSC.

Multifunctionality

B-19. The MH sections have a priority of prevention activities, but also have the capability to conduct or support any combat operational stress function as required. The COSC teams will be outfitted to a standard to gain the capability of prevention and limited fitness activities.

Modularity, Command and Control

B-20. The medical detachment, COSC attached early to supported units will have a better-integrated relationship. Detachment commanders and NCOICs will collocate with and coordinate all activities through supported unit surgeons. Detachments are designed to support a division in combat or a joint task force of up to 25,000 personnel in noncombat operations. The BH teams are designed to support and operate within maneuver BCTs in support of maneuver battalions for area coverage, collocating in BSA with the BSMC. Teams may be combined for task-organization to meet the entire spectrum of COSC as required.

Limitations

B-21. The medical detachment, COSC does not have its own logistics or services capability and must depend on the supported unit for logistical, legal, FHP, food service, finance, and personnel and administrative services. The COSC teams are small (one or two vehicles) and are dependent upon supported units for convoy and security operations. The logistical support necessary (space, tents, cots, and so forth) to provide Soldier restoration/reconstitution functions must be furnished by the supported element (division, corps, or theater).

Assumptions

B-22. Wartime activities will continue to cause physical and mental impacts of extremely high stress in both direct combat and the support operational environment. Advances in technology will impact the individual Soldier by increasing dispersion, and adding to their isolation and stress levels. The enemy of the future will look for methods that will have the greatest psychological impact on our future Soldier; future enemies will not be our technological equals and will increasingly rely on terror and nontraditional methods to unnerve, injure, and demoralize both Soldiers and civilians. Commanders will need to be able to quickly and effectively apply COSC capabilities at every level of the unit.

Appendix C

Medical Company and Medical Detachment, Combat Stress Control (Medical Force 2000)

SECTION I — MEDICAL COMPANY, COMBAT STRESS CONTROL (TOE 08467L000) (MEDICAL FORCE 2000)

MISSION

C-1. The mission of the AOE medical company, CSC is to provide comprehensive COSC support through directed interventions activities and COSC training for supported corps units. It provides DS to maneuver brigades lacking organic BH officers; augments units with BH assets; and provides area support. The medical company, CSC reconstitutes other COSC assets. The medical company, CSC provides COSC interventions and activities to indigenous populations as directed in stability and reconstruction operations, to include domestic support operations, humanitarian assistance, disaster relief, peace support operations, and detention facility operations. The medical company, CSC provides COSC interventions and activities to units in support of their readiness preparation and throughout their deployment cycle.

ASSIGNMENT

C-2. The AOE medical company, CSC (TOE 08467L000) is assigned to a corps MEDBDE (TOE 08422A100) or corps MEDCOM (TOE 08411A000).

EMPLOYMENT

C-3. Company headquarters usually locates with the corps medical headquarters. The employment of its teams includes their dispersal throughout the corps AO. Teams may be attached to an ASMB; a division or brigade medical company; a CSH; or other corps medical unit headquarters.

CAPABILITIES

C-4. At TOE Level 1, the medical company, CSC includes—

- Dividing the preventive section into six mobile COSC preventive teams.
- Dividing the restoration section into four mobile restoration teams with each team being equipped to hold 40 Soldiers at the same time. With additional logistical support, each team can accommodate additional Soldiers under surge conditions.
- Assisting higher headquarters BH staff with planning and coordination of COSC support, identifying the stress threat and mental and physical stressors, and the implementation of COSC functional areas.

DEPENDENCY

C-5. This unit is dependent on—

- Appropriate elements of the corps for legal; finance; field feeding; personnel and administrative services support; laundry and clothing exchange; mortuary affairs support; and security of EPW, detainee, and US prisoner patients.
- The medical headquarters to which it is assigned/attached for FHP; medical administration; logistics (including MEDLOG); medical regulating of patients; evacuation; coordination for RTD; and unit-level equipment and CE maintenance.

Mobility and Security

C-6. This unit is 100-percent mobile. It requires 100 percent of its organic personnel and equipment be transported in a single lift, using its organic vehicles. Upon relocation, Soldiers being held will require additional transportation. This unit is responsible for perimeter defense of its immediate operational area. However, it is dependent on appropriate elements of the corps for additional security, to include security of convoy operations. Personnel of the company (except the chaplain) are provided weapons for their personal defense and for the defense of their patients and/or held Soldiers.

Organization

C-7. The AOE medical company, CSC (Figure C-1) is organized into a headquarters section, a preventive section with six (modular) CSCP, and a restoration section with four (modular) combat stress control restoration (CSCR) teams. The medical company, CSC is only present in the MF2K and is an RC organization that is programmed to be in the Army inventory until the year 2010. Under the MRI a new medical detachment, CSC, TOE 08463A000, will replace this unit.

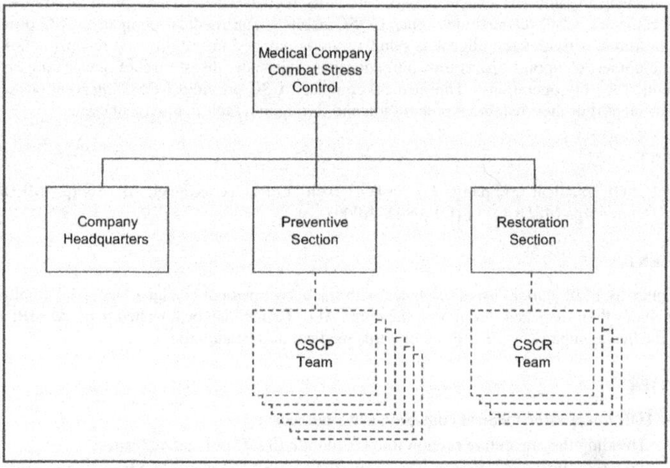

Figure C-1. Medical company, combat stress control

Company Headquarters Section

C-8. The company headquarters section provides C2 and unit-level administrative and maintenance support to its subordinate sections when they are collocated with the company. Personnel from the headquarters section are deployed with teams or task-organized COSC elements, as required. Personnel assigned to this section are identified in Table C-1.

Table C-1. Company headquarters section personnel 3/0/14

Company Commander (LTC/05, AOC 60W00, MC)
Chaplain (CPT/03, AOC 56A00, CH)
Field Medical Assistant (CPT/03, AOC 70B67, MS)
First Sergeant (1SG/E8, MOS 68W5M)
Mental Health NCO (SFC/E7, MOS 68X40)
Motor Sergeant (SSG/E6, MOS 63B30)
Supply Sergeant (SGT/E5, MOS 92Y20)
Equipment Record/Parts Sergeant (SGT/E5, MOS 92A20)
Wheeled Vehicle Mechanic (SGT/E5, MOS 63B20)
Human Resource Specialist (SPC/E4, MOS 42A10)
Nuclear, Biological, and Chemical Specialist (SPC/E4, MOS 74D10)
Wheeled Vehicle Mechanic (SPC/E4, MOS 63B10)
Unit Supply Specialist/Armorer (SPC/E4, MOS 92Y10)
Power-Generation Equipment Repairer (SPC/E4, MOS 52D10)
Cook (Three)— (Two SPC/E4 and One PFC/E3, MOS 92G10)
Note. The cook is deployed to assist the unit providing food service support to the company. He may participate in other CSC unit missions including outreach surveillance and stress control training.

Preventive Section

C-9. The 24-person preventive section staff is identified in Table C-2. The preventive section is task-organized to conduct COSC interventions and activities. Preventive section personnel may be task-organized with personnel of the COSC restoration section into teams for specific missions. The preventive section can divide into six CSCP teams. The section (and team) leader position may be held by any of the officers assigned to the section. The preventive section can augment or reconstitute the medical detachment, CSC teams.

Table C-2. Preventive section personnel 12/0/12

Psychiatrist (Three)—(MAJ/04, AOC 60W00)
Social Worker (Two)—(MAJ/04, AOC 73A67)
Psychiatrist (Three)—(CPT/03, AOC 60W00)
Social Worker (Four)—(CPT/03, AOC 73A67)
Team Chief (Six)—(SGT/E5, MOS 68X20)
Mental Health Specialist (Six)—(SPC/E4, MOS 68X10)

Restoration Section

C-10. The 44-person restoration section staff is identified in Table C-3. The restoration section may be task-organized to perform its COSC mission. Combat stress control restoration section personnel may also be task-organized with personnel of the CSCP section into teams for specific missions. This section can be divided into four CSCR teams. Each CSCR team can deploy a four-person mobile team using their HMMWV.

Table C-3. Restoration section personnel 12/0/32

Occupational Therapist (MAJ/04, AOC 65A00)
Psychiatric/Mental Health Nurse (Two)—(MAJ/04, AOC 66C00/7T)
Clinical Psychologist (MAJ/04, AOC 73B67)
Occupational Therapist (Three)—(CPT/03, AOC 65A00)
Psychiatric/Mental Health Nurse (Two)—(CPT/03, AOC 66C00/7T)
Clinical Psychologist (Three)—(CPT/03, AOC 73B67)
Mental Health NCO (Two)—(SFC/E7, MOS 68X40)
Occupational Therapy NCO (SSG/E6, MOS 68W30/N3)
Team Chief (Four)—(SSG/E6, MOS 68X30)
Patient Administration NCO (SGT/E5, MOS 68G20)
Occupational Therapy Sergeant (Two)—(SGT/E5, MOS 68W20/N3)
Mental Health NCO (Four)—(SGT/E5, MOS 68X20)
Patient Administration Specialist (Three)—(Two SPC/E4 and One PFC/E3, MOS 68G10)
Occupational Therapy Specialist (Five)—(SPC/E4, MOS 68W10/N3)
Mental Health Specialist (Ten)—(Six SPC/E4 and Four SPC/E3, MOS 68X10)

SECTION II — MEDICAL DETACHMENT, COMBAT STRESS CONTROL (TOE 08463L000) (MEDICAL FORCE 2000)

MISSION

C-11. The mission of the AOE MF2K medical detachment, CSC is to provide COSC interventions and activities to supported units in its AO. It augments division and brigade MH sections; provides direct support to combat brigades without organic BH officers; and provides area support in its AO. The medical detachment, CSC reconstitutes other brigade and division COSC assets. The medical detachment, CSC provides COSC interventions and activities to indigenous populations as directed in stability and reconstruction operations, to include domestic support operations, humanitarian assistance, disaster relief, peace support operations, and detention facility operations. The medical detachment, CSC provides COSC interventions and activities between deployments to units in support of their readiness preparation and throughout their deployment cycle. Under MRI this detachment will be replaced with medical detachment, CSC, TOE 08463A000.

> ***Note.*** The MF2K and MRI medical detachments, CSC conduct COSC support operations very similarly. However, in the MRI medical detachment, CSC, psychiatry assets are assigned to the restoration section and psychology assets are assigned to the preventive section whereas in the MF2K detachment, the manning strategy for psychology and psychiatry is reversed: psychologists are in the restoration section and psychiatrists are on the preventive teams.

ASSIGNMENT

C-12. The MF2K medical detachment, CSC (TOE 08463L000), is assigned to a corps MEDBDE (TOE 08422A100), or a corps MEDCOM (TOE 08411A000), or other medical task force C2 elements. Its teams may be attached to a medical company, CSC (TOE 08467L000); an ASMB (TOE 08456A000); a brigade or divisional medical company; or a CSH.

EMPLOYMENT

C-13. Detachment headquarters is usually located in the DSA. Its teams disperse and are employed throughout its AO.

CAPABILITIES

C-14. The MF2K medical detachment, CSC provides—

- Mobile COSC interventions from the preventive section which can divide into three preventive teams.
- Holding for 40 Soldiers by the restoration team for COSC interventions such as restoration. With additional logistical support, the team can accommodate additional Soldiers under surge conditions.

STAFF RESPONSIBILITIES

C-15. The medical detachment assists the C2 headquarters (to which it is assigned or attached) regarding planning and coordination of COSC support, stress threat, mental and physical stressors, stress behaviors, principles of COSC, and implementation of COSC functional areas.

DEPENDENCY

C-16. This unit is dependent on—

- Appropriate elements of the supporting unit for FHP; religious support; legal; finance; field feeding; personnel and administrative services support; laundry and clothing exchange; mortuary affairs support; and security of EPW, detainee, and US prisoner patients.
- Supporting unit for medical administration; logistics (including MEDLOG); medical regulating; evacuation; coordination for RTD; and unit-level equipment and CE maintenance.

MOBILITY AND SECURITY

C-17. This unit is 100-percent mobile. It requires 100 percent of its organic personnel and equipment be transported in a single lift, using its organic vehicles. Upon relocation, Soldiers being held will require additional transportation. This unit is responsible for perimeter defense of its immediate operational area. However, it is dependent on appropriate elements of the supporting unit for additional security, to include security of convoy operations. Personnel of the detachment are provided weapons for their personal defense and for the defense of their patients and/or held Soldiers.

ORGANIZATION

C-18. This 25-person unit (see Table C-4) is composed of a headquarters, a combat stress preventive section with three preventive teams, and a CSCR. The modular CSC teams found in the MF2K CSC medical detachment are similar to those found in the CSC medical company.

Table C-4. Detachment teams

Headquarters	*Preventive Section*	*Restoration Section*
	CSCP#1	CSCR
	CSCP#2	
	CSCP#3	

C-19. The detachment headquarters provides C2 for the detachment. It is responsible for planning, coordinating, and implementing COSC interventions and activities for supported units. It consists of three personnel: a detachment Commander (05, 60W00, MC), a detachment sergeant (E7, 68X40), and a wheeled-vehicle mechanic (E4, 63B10). The detachment commander also serves as a treating physician with the preventive section. The detachment NCOIC (a senior MH NCO) also serves as the restoration team sergeant. Detachment officers and NCOs from the prevention team and the restoration team may be assigned additional duties, which enhance the overall effectiveness of the headquarters section. See Table C-5 for personnel assigned to detachment headquarters.

Table C-5 Detachment headquarters personnel

Commander (LTC/05, AOC 60W00)
Detachment Sergeant (SFC/E7, MOS 68X40)
Wheeled Vehicle Mechanic (SPC/E4, MOS 63B10)
Note. The commander also serves as the psychiatrist in the preventive section.

Preventive Section

C-20. The 12-person preventive section staff is identified in Table C-6. The preventive section may be task-organized to conduct its COSC mission. Preventive section personnel may also be task-organized with personnel of the restoration section into teams for specific missions. The preventive section can divide into three preventive teams. The section (and team) leader position may be held by any of the officers assigned to the section. This section's COSC interventions and activities are—

- Unit needs assessment; consultation and education; critical event and transition management; COSC triage; stabilization (emergency); and BH treatment.
- Assisting with restoration and reconditioning at the CSC detachment program. Overseeing a 1- to 3-day COSC restoration program in a brigade, division, or ASMC holding section or in another area suitable for Soldiers experiencing COSR and/or other stress-related disorders.

Table C-6. Preventive section 7/0/6

Psychiatrist (LTC/05, AOC 60W00)
Clinical Psychologist (Three)—(MAJ/04, AOC 73B67)
Social Worker (Three)—(CPT/03, AOC 73A67)
Team Chief (Three)—(SGT/E5, MOS 68X20)
Mental Health Specialist (Three)—(SPC/E4, MOS 68X10)
Note. The psychiatrist is also counted in the headquarters section.

Restoration Team

C-21. The 9-person restoration team staff is identified in Table C-7. The restoration team is task-organized to provide COSC interventions and activities. Restoration team personnel may also be task-organized with personnel of the CSC preventive section into teams for specific missions. Each restoration team can deploy a four-person mobile team with a HMMWV. The CSCF usually collocates with a supported divisional medical company to provide mobile COSC support within a DSA and conduct restoration programs, as required. The CSCF provides staff and equipment for operating a restoration or reconditioning center. This section's COSC interventions and activities are—

- Unit needs assessment; consultation and education; critical event and transition management; COSC triage; stabilization; and BH treatment.
- Conducting Soldier restoration and reconditioning programs.
- Assisting the CSH psychiatric section when psychiatric ward capability is required (refer to FM 8-10-14).

Table C-7. Restoration team 3/0/6

Psychiatrist (MAJ/04, AOC 60W00)
Psychiatric/Mental Health Nurse (MAJ/04, AOC 66COO/7T)
Occupational Therapist (CPT/03, AOC 65A00)
Occupational Therapy NCO (SSG/E6, 68W30/N3)
Team Chief (SSG/E6, MOS 68X30)
Occupational Therapy Sergeant (SGT/E5, MOS 68W20/N3)
Mental Health NCO (SGT/E5, MOS 68X20)
Mental Health Specialist (Two)—(One SPC/E4 and One PFC/E3, MOS 68X10)

Glossary

SECTION I — ACRONYMS AND ABBREVIATIONS

AAR	after-action review
ABCA	American, British, Canadian, and Australian
ACR	armored cavalry regiment
AMEDD	Army Medical Department
AN	Army Nurse Corps
AO	area of operations
AOC	area of concentration
AOE	Army of Excellence
AR	Army regulation
ART	Army tactical task
ASMB	area support medical battalion
ASMC	area support medical company
AUTL	Army Universal Task List
BAS	battalion aid station
BCT	brigade combat team
BDP	behavioral disordered patient
BF	battle fatigue
BH	behavioral health
BICEPS	brevity, immediacy, contact, expectancy, proximity, and simplicity (See Section II for definition.)
BSA	brigade support area
BSB	brigade support battalion
BSMC	brigade support medical company
C	Centigrade
C2	command and control
C4I	command, control, communications, computers, and intelligence
CBRN	chemical, biological, radiological, and nuclear
CE	communications-electronics
CH	chaplain
CHL	combat health logistics
CHS	combat health support
COA	course(s) of action
CONUS	continental United States
COSC	combat and operational stress control (See Section II for definition.)
COSR	combat and operational stress reaction (See Section II for definition.)

CP	command post
CPT	captain
CRST	combat religious support team
CS	combat support
CSC	combat stress control
CSCF	combat stress control fitness
CSCP	combat stress control preventive
CSCR	combat stress control restoration
CSH	combat support hospital
CSR	combat stress reaction
CZ	combat zone
DA	Department of the Army
DD	Department of Defense
DEPMEDS	Deployable Medical Systems
DISCOM	division support command
DMHS	division mental health section
DNBI	disease and nonbattle injury(ies)
DOD	Department of Defense
DODD	Department of Defense Directive
DODI	Department of Defense Instruction
DS	direct support
DSA	division support area
EAB	echelons above brigade
EAC	echelons above corps
EAD	echelons above division
EMT	emergency medical treatment
EPW	enemy prisoner(s) of war
ETOD	end-of-tour debriefing
F	Fahrenheit
FAP	Family Advocacy Program
FHP	force health protection
1LT	first lieutenant
1SG	first sergeant
FM	field manual
FMC	United States Field Medical Card (Department of Defense Form 1380)
FRAGO	fragmentary order
FSB	forward support battalion
FSMC	forward support medical company
G-1	Assistant Chief of Staff (Personnel)
G-2	Assistant Chief of Staff (Intelligence)
G-3	Assistant Chief of Staff (Operations and Plans)

G-4	Assistant Chief of Staff (Logistics)
G-5	Assistant Chief of Staff (Plans and Policy)
G-9	Assistant Chief of Staff (Civil-Military Affairs)
GP	general purpose
HBCT	heavy brigade combat team
HHD	headquarters and headquarters detachment
HIP	help-in-place
HMMWV	high-mobility multipurpose wheeled vehicle
HSL	health service logistics
HSS	health service support
IED	improvised explosive device
JAG	Judge Advocate General
LTC	lieutenant colonel
MAJ	major
MC	Medical Corps
MCW	minimal care ward
MEDBDE	medical brigade
MEDCEN	medical center
MEDCOM	medical command (US Army)
MEDDAC	medical department activity
MEDLOG	medical logistics
METT-TC	mission, enemy, terrain and weather, troops and support available, time available, and civil considerations
MF2K	Medical Force 2000
mg	milligram(s)
MH	mental health
MHAT	Mental Health Advisory Team (US Army)
MMB	multifunctional medical battalion
MOS	military occupational specialty
MP	military police
MRI	Medical Reengineering Initiative
MRO	medical regulating officer
MS	Medical Service Corps
MSB	main support battalion
MSMC	main support medical company
MTF	medical treatment facility
MTOE	modification table of organization and equipment
NATO	North Atlantic Treaty Organization
NCO	noncommissioned officer
NCOIC	noncommissioned officer in charge
NP	neuropsychiatric

OEF	Operation Enduring Freedom
OIF	Operation Iraqi Freedom
OPCON	operational control
OPLAN	operation plan
OPORD	operation order
OPTEMPO	operational tempo
OSAT	operational stress assessment team
OSR	operational stress reaction
OT	occupational therapy
PA	physician assistant
PCP	phencyclidine hydrochloride
PD	psychological debriefing
PERSTEMPO	personnel tempo
PIES	proximity, immediacy, expectancy, and simplicity
POC	point of contact
PTE	potentially traumatizing event
PTSD	posttraumatic stress disorder
pub	publication
PVNTMED	preventive medicine
R&R	rest and relaxation
RC	Reserve Component
RTD	return to duty
S1	Adjutant (US Army)
S2	Intelligence Officer (US Army)
S3	Operations and Training Officer (US Army)
S4	Supply Officer (US Army)
SBCT	Stryker Brigade Combat Team
SGT	sergeant
SOI	signal operation instructions
SOP	standing operating procedure
SP	Army Medical Specialty Corps
SPC	specialist
SRC	standard requirement code
SSG	staff sergeant
TC	training circular
TDA	table of distribution and allowances
TEM	traumatic events management
TEMPER	tent, expandable, modular, personnel
TG	technical guide
TOE	table(s) of organization and equipment
TRADOC	United States Army Training and Doctrine Command

TSOP	tactical standing operating procedures
UCMJ	Uniform Code of Military Justice
UMT	unit ministry team
UNA	unit needs assessment
US	United States
USAMEDDC&S	United States Army Medical Department Center and School
USAMRMC	United States Army Medical Research and Materiel Command
USAR	United States Army Reserve
USMC	United States Marine Corps
VHA	Veterans Health Adminstration

SECTION II — TERMS

brevity, immediacy, contact, expectancy, proximity, and simplicity

An acronym used for the management of combat and operational stress reactions—*brevity* (usually less than 72 hours); *immediacy* (as soon as symptoms are evident); *contact* (chain of command remains directly involved in the Soldiers recovery and return to duty), *expectancy* (combat stress control unit personnel expectation that casualties will recover); *proximity* (of treatment at or as near the front as possible); *simplicity* (the use of simple measures such as rest, food, hygiene, and reassurance). Also known as BICEPS.

combat and operational stress control

A coordinated program for the prevention of and actions taken by military leadership to prevent, identify, and manage adverse combat and operational stress reactions in units. Also known as COSC.

combat and operational stress reaction

The expected, predictable, emotional, intellectual, physical, and/or behavioral reactions of Service members who have been exposed to stressful events in combat or military operations other than war. Also known as COSR.

5 R's

Actions used for combat and operational stress reaction control that include—**R**eassure of normality; **R**est (respite from combat or break from the work); **R**eplenish bodily needs (such as thermal comfort, water, food, hygiene, and sleep); **R**estore confidence with purposeful activities and contact with his unit; **R**eturn to duty and reunite Soldier with his unit.

reconditioning program

An intensive 4- to 7-day program (may be extended by exception to theater evacuation policy) of replenishment, physical activity, therapy, and military retraining for combat and operational stress control casualties and neuropsychiatric cases (including alcohol and drug abuse) who require successful completion for return to duty or is evacuated for further neuropsychiatric evaluation.

Soldier restoration

A 24- to 72-hour (1- to 3-day) program in which Soldiers with combat and operational stress reactions receive treatment.

stabilization

The initial short-term management and evaluation of severely behaviorally disturbed Soldiers caused by an underlying combat and operational stress reaction, behavioral health disorder, or alcohol and/or drug abuse reaction.

References

SOURCES USED

These are the sources quoted or paraphrased in this publication.

DEPARTMENT OF DEFENSE

These documents are available online at: http://www.dtic.mil/whs/directives/

DODD 6490.1, *Mental Health Evaluations of Members of the Armed Forces*, 1 October 1997

DODD 6490.2, *Comprehensive Health Surveillance*, 21 October 2004

DODD 6490.5, *Combat Stress Control (CSC) Program*, 23 February 1999

DODI 6490.4, *Requirements for Mental Health Evaluations of Members of the Armed Forces,* 28 August 1997

MULTISERVICE PUBLICATION

FM 6-22.5/MCRP 6-11C/NTTP 1-15M, *Combat Stress*, 23 June 2000

ARMY PUBLICATIONS

These publications are available online at: https://akocomm.us.army.mil/usapa

AR 40-216, *Neuropsychiatry and Mental Health,* 10 August 1984

FM 7-15, *The Army Universal Task List*, 31 August 2003

FM 22-51, *Leaders' Manual for Combat Stress Control,* 29 September 1994

DOCUMENTS NEEDED

These documents must be available to the intended users of this publication.

DEPARTMENT OF DEFENSE FORMS

These forms are available online at: https://akocomm.us.army.mil/usapa

DD Form 1380, *US Field Medical Card*

DD Form 2795, *Pre-Deployment Health Assessment*

DD Form 2796, *Post-Deployment Health Assessment*

MULTISERVICE PUBLICATION

AR 190-8/OPNAVINST 3461.6/AFJI 31-304/MCO 3461.1, *Enemy Prisoners of War, Retained Personnel, Civilian Internees, and Other Detainees,* 1 October 1997

ARMY PUBLICATIONS

These publications are available online at: https://akocomm.us.army.mil/usapa

AR 25-1, *Army Knowledge Management and Information Technology Management*, 15 July 2005

AR 40-3, *Medical, Dental, and Veterinary Care,* 12 November 2002

AR 40-61, *Medical Logistics Policies*, 28 January 2005

AR 40-66, *Medical Record Administration and Health Care Documentation,* 20 July 2004

AR 40-68, *Clinical Quality Management,* 26 February 2004

AR 40-400, *Patient Administration,* 12 March 2001

AR 40-501, *Standards of Medical Fitness,* 1 February 2005

AR 340-21, *The Army Privacy Program,* 5 July 1985

DA Pamphlet 611-21, *Military Occupational Classification and Structure,* 31 March 1999

FM 4-02, *Force Health Protection in a Global Environment,* 13 February 2003

FM 4-02.6, *The Medical Company—Tactics, Techniques, and Procedures,* 1 August 2002 (Change 1, 9 April 2004)

FM 4-02.7, *Health Service Support in a Nuclear, Biological, and Chemical Environment—Tactics, Techniques, and Procedures,* 1 October 2002

FM 4-02.10, *Theater Hospitalization,* 3 January 2005

FM 4-02.16, *Army Medical Information Management—Tactics, Techniques, and Procedures,* 22 August 2003

FM 4-02.17, *Preventive Medicine Services,* 28 August 2000

FM 4-25.12, *Unit Field Sanitation Team,* 25 January 2002

FM 8-10-6, *Medical Evacuation in a Theater of Operations—Tactics, Techniques, and Procedures,* 14 April 2000

READINGS RECOMMENDED

These sources contain relevant supplemental information.

JOINT OR MULTISERVICE PUBLICATIONS

Most joint publications are available online at: http://www.dtic.mil/doctrine/

Joint Pub (JP) 1-02, *Department of Defense Dictionary of Military and Associated Terms,* 12 April 2001 (As amended through 9 June 2004)

JP 3-0, *Doctrine for Joint Operations,* 9 September 2001

JP 4-02, *Doctrine for Health Service Support in Joint Operations,* 30 July 2001

FM 1-02/MCRP 5-12A, *Operational Terms and Graphic,* 21 September 2004

FM 4-25.11/NTRP 4-02.1.1/AFMAN 44-163(I)/MCRP 3-02G), *First Aid,* 23 December 2002 (Change 1, 15 July 2004)

FM 8-9/NAVMED P-5059/AFJMAN 44-151V1V2V3, *NATO Handbook on the Medical Aspects of NBC Defensive Operations AMedP-6(B), Part I—Nuclear, Part II—Biological, Part IIII—Chemical,* 1 February 1996

FM 8-284/NTRP 4-02.23(NAVMED P-5042)/AFMAN (I) 44-156/MCRP 4-11.1C, *Treatment of Biological Warfare Agent Casualties,* 17 July 2000 (Change 1, 8 July 2002)

FM 8-285(4-02.285)/NAVMED P-5041/AFJMAN 44-149/FMFM 11-11, *Treatment of Chemical Agent Casualties and Conventional Military Chemical Injuries,* 22 December 1995

FM 21-10/MCRP 4-11.1D, *Field Hygiene and Sanitation,* 21 June 2000

ARMY PUBLICATIONS

Most of these publications are available online at: https://akocomm.us.army.mil/usapa

AR 71-32, *Force Development and Documentation—Consolidated Policies*, 3 March 1997

AR 635-200, *Active Duty Enlisted Administrative Separations*, 6 June 2005

FM 3-0, *Operations*, 14 June 2001

FM 3-06.11, *Combined Arms Operations in Urban Terrain*, 28 February 2002

FM 3-07, *Stability Operations and Support Operations*, 20 February 2003

FM 3-19.40, *Military Police Internment/Resettlement Operations*, 1 August 2001

FM 3-21.21, *The Stryker Brigade Combat Team Infantry Battalion*, 8 April 2003 (Change 1, 31 July 2003)

FM 3-21.31, *The Stryker Brigade Combat Team*, 13 March 2003

FM 3-90.3, *The Mounted Brigade Combat Team*, 1 November 2001

FM 5-0, *Army Planning and Orders Production*, 20 January 2005

FM 4-0, *Combat Service Support*, 29 August 2003

FM 4-02.4, *Medical Platoon Leaders' Handbook—Tactics, Techniques, and Procedures*, 24 August 2001 (Change 1, 18 December 2003)

FM 4-02.19, *Dental Service Support in the Theater of Operations*, 1 March 2001

FM 4-02.21, *Division and Brigade Surgeons' Handbook (Digitized)—Tactics, Techniques, and Procedures*, 15 November 2000

FM 4-02.24, *Area Support Medical Battalion— Tactics, Techniques, and Procedures*, 28 August 2000

FM 4.25.10, *Field Hygiene and Sanitation*, 20 June 2002

FM 8-10-14, *Employment of the Combat Support Hospital Tactics, Techniques, and Procedures*, 29 December 1994 (will be revised as FM 4-02.14)

FM 8-42, *Combat Health Support in Stability Operations and Support Operations*, 27 October 1997 (will be revised as FM 4-02.42)

FM 8-50, *Prevention and Medical Management of Laser Injuries*, 8 August 1990 (will be revised as FM 4-02.50)

FM 8-55, *Planning for Health Service Support,* 9 September 1994 (will be revised as FM 4-02.55)

FM 10-27-4, *Organizational Supply and Services for Unit Leaders,* 14 April 2000 (will be revised as FM 4-20.05)

FM 27-1, *Legal Guide for Commanders.* 13 January 1992 (will be revised as FM 1-04.1)

FM 27-10, *The Law of Land Warfare,* 18 July 1956 (Reprinted with basic including Change 1, 15 July 1976) (will be revised as FM 1-04.10)

FM 63-20, *Forward Support Battalion*, 26 February 1990 (will be revised as FM 4-93.20)

FM 63-21, *Main Support Battalion*, 7 August 1990 (will be revised as FM 4-93.21)

FM 100-9, *Reconstitution*, 13 January 1992 (will be revised as FM 4-100.9)

FM 100-17, *Mobilization, Deployment, Redeployment, and Demobilization,* 28 October 1992 (will be revised as FM 3-35)

FM 100-17-5, *Redeployment*, 29 September 1999 (will be revised as FM 3-35)

MEDCOM Regulation 40-38, *Command-Directed Mental Health Evaluation*, 1 September 2001

TC 25-20, *A Leader's Guide to After-Action Reviews,* 30 September 1993

TRADOC Pamphlet 600-22, *Leaders Guide for Suicide Prevention Planning,* 16 February 2005 (Available at http://www.tradoc.army.mil/publications.htm)

United States Army Center for Health Promotion and Preventive Medicine

Most of these publications are available online at: http://chppm-www.apgea.army.mil/dhpw/

USACHPPM TG 240, *Combat Stress Behaviors,* June 2004

USACHPPM TG 241, *Combat Operational Stress Reaction* (*COSR*) (*"Battle Fatigue"*), June 2004

USACHPPM TG 242, *Combat Operational Stress Reactions* (*COSR*) *Prevention: Leader Actions,* June 2004

USACHPPM TG 243, *Combat Stress Card,* May 1999

Tables of Organization and Equipment

TOE 08411A000, *Medical Command, Corps* (*MRI*)

TOE 8422A100, *Medical Brigade, Corps* (*MRI*)

TOE 08456A000, *Area Support Medical Battalion*

TOE 08457A000, *Area Support Medical Company*

TOE 08463A000, *Medical Detachment, Combat Stress Control* (*MRI*)

TOE 08463L000, *Medical Detachment, Combat Stress Control* (*MF2K*)

TOE 08467L000, *Medical Company, Combat Stress Control* (*MF2K*)

TOE 08660G000, *Medical Detachment, Combat Operational Stress Control* (*Force Development Update*)

Department of Veterans Affairs

Veterans Health Administration, *Clinical Practice Guidelines (Mental Health),* available online at: http://www.oqp.med.va.gov/cpg

Index

References are to paragraph numbers except where otherwise specified.

TABLE OF CONTENTS

PART I. SCOUTING

PART I. SCOUTING

CHAPTER 1. FUNDAMENTALS OF SCOUTING

Scouting involves observing terrain and/or the enemy, and accurately reporting those observations. Scouting requires proficiency in the use of weapons, cover and concealment, route selection, and skill in unobserved day or night movement.

1001. PURPOSE

When an infantry unit is not actively fighting the enemy, it should be actively searching for the enemy, which is the primary purpose of scouting. The unit attempts to keep the enemy off balance while making preparations for further attacks. Physically locating and keeping the enemy off balance are normally accomplished by small units ranging from a two-person scouting party to a squad-size patrol.

Infantrymen are sent out as scouts or as members of a patrol because the commander needs information about the enemy, terrain, and the location of friendly troops. The lives of the entire unit may depend upon the success or failure of a scout or patrol and the accuracy and timeliness of the report. The success of the scout or patrol will depend upon their training, preparation by the commander, and understanding their mission and the commander's requirements.

To wage combat successfully, a commander must have accurate, detailed, and timely information about the enemy, the terrain, and adjacent friendly units. Well-trained scouts and capably led patrols are among the most effective means the commander has for acquiring the information necessary to plan tactical actions and make decisions in execution.

1002. REQUIRED SCOUTING SKILLS

To be effective, a scout must be able to—

- Recognize terrain features.
- Read a map and determine direction.
- Practice and implement the principles of cover and concealment.
- Fully utilize movement and route selection.
- Know the enemy (estimate enemy unit composition and strength).
- Observe and report information accurately.
- Select routes and move through numerous types of terrain.

CHAPTER 2. TERRAIN, MAPS, AND DIRECTION

A scout must understand map symbols, identify elevations from contour interval lines, scale distance on a map, relate natural and man-made features shown on the map to the actual features on the ground, plot a course from one point to another, and locate his current position. To relate a map to the actual terrain and its features, a scout must be able to orient it to the ground using a compass, two points, a watch, and the sun or the stars.

2001. TERRAIN FEATURES

Since the infantry works and fights on the ground, terrain ashore information that scouts gather and report on is of great importance. Hills, valleys, woods, and streams are the forms and growths commonly referred to as natural land features. Artificial or man-made features include houses, bridges, and railroads. Figure 2-1 shows some important terrain features.

2002. THE LENSATIC COMPASS

The best method of finding direction, during both day and night, is with a compass. The lensatic and M2 are the two types of compasses issued to infantry units. Both work on the same principle.

The standard compass for general use in the Marine Corps is the pivot-mounted lensatic compass, so called because azimuths are read through a magnifying lens in the eyepiece. Figure 2-2, on page 2-2, shows the lensatic compass and its nomenclature. The meter graphic scale on the side of the compass is 1:50,000, which is the most used scale in military mapping. The graphic scale is useful in the field as a straightedge, an aid in orienting the map, and a means of reading map grid coordinates. The plastic dial is graduated in both degrees and mils. Numbers on the dial are printed in black. There are luminous markings on the bezel, floating dial, and on both ends of the sighting wire, plus a 3-degree bezel serration and clicking device that permit reading azimuths at night. The compass is carried in a nylon case that may be attached to the cartridge belt.

Magnetic compasses are affected by the presence of iron or magnetic fields. Consequently, the scout should not be within the influence of local magnetic

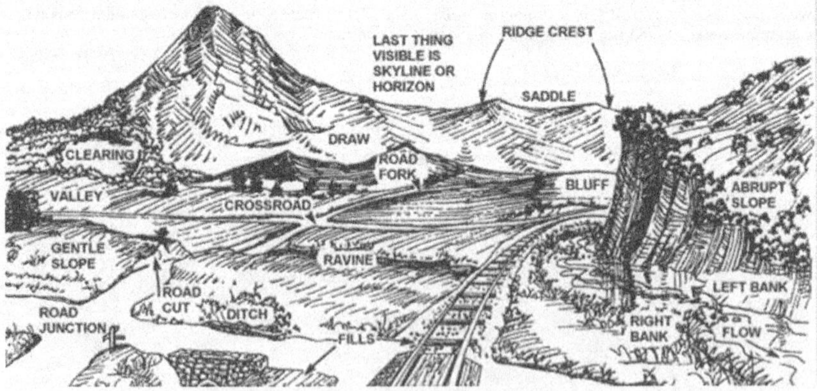

Figure 2-1. Natural and Artificial Terrain Features.

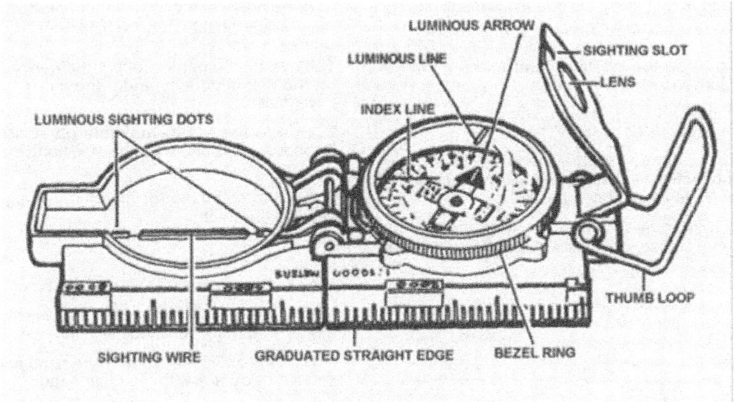

Figure 2-2. Lensatic Compass.

attraction while using a compass to determine direction. The rifle, pistol, and other metal objects must be laid aside when reading the compass. The minimum distances the scout should be from visible masses of iron and electrical fields of magnetism for the compass to provide accurate readings are provided below.

Magnetic Fields	Minimum Distances (meters)
High tension powerlines	60
Field guns	20
Vehicles (wheeled or tracked)	20
Telephone and telegraph wires	10
Barbed wire	10
Machine gun	3
Rifle, pistol	1

Reading an Azimuth

To read an azimuth to any point, the cover of the compass is raised to an angle of 90 degrees in relation to the index face, and the eyepiece is lifted to a 45-degree angle in relation to the bezel, or so the numbers on the dial can be seen. The thumb of either hand is placed in the thumb loop, the index finger extended along the side of the compass case, and the remainder of the hand closed. The closed hand and wrist are grasped with the other hand. The elbows are drawn in close to the body, forming a firm foundation for the compass. The eye is placed to the lens on the eyepiece. The compass is pointed at the object or point to which the azimuth is to be read. A sighting is taken through the sighting slot in the eyepiece, and the point is lined up with the sighting wire in the cover. The compass is held until the dial steadies; then the reading is taken through the lens of the eyepiece. This reading is the magnetic azimuth of the line from the observer to the point.

Reading a Back Azimuth

A back azimuth is the direction opposite the line of sight. If the azimuth is less than 180 degrees, the back azimuth is obtained by adding 180 degrees. If the azimuth is greater than 180 degrees, the back azimuth is obtained by subtracting 180 degrees. Back azimuths are used to determine a return route or to resection to determine a current position.

Circumventing Obstacles

When a scout is traveling on an azimuth and comes upon an obstacle—such as a contaminated area, minefield or swamp—the following steps (sometimes referred to as the *90-degree offset* method) is employed to go around or circumvent the obstacle and resume movement along the original azimuth (see fig. 2-3). The steps are as follows:

1. Move up to the obstacle and make a full 90-degree turn to the right (or left).

2. Walk beyond the obstacle, keeping track of the distance in paces or meters.

3. Stand at the end of the obstacle, face in the original direction of march, and follow that azimuth until the obstacle has been passed.

4. Make a 90-degree turn to the left (or right) and move the distance previously measured to return to the original line of march.

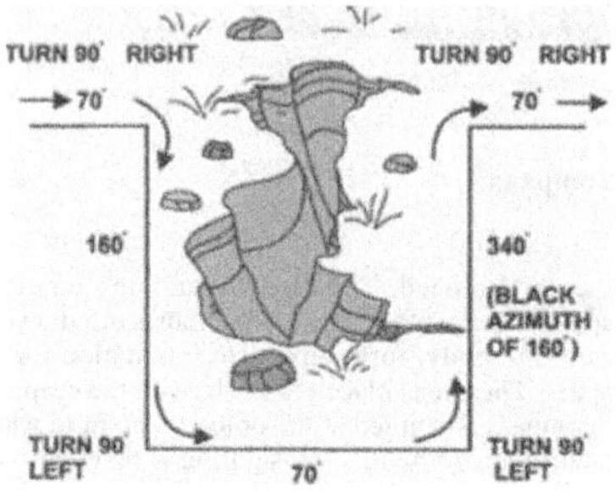

Figure 2-3. Circumventing an Obstacle.

Following an Azimuth During the Day

The eye is placed to the lens in the eyepiece and the compass moved until the desired azimuth reading is visible beneath the fixed index. Without moving the compass, the vision is shifted from the lens through the sighting slot in the eyepiece, and a sighting is taken out beyond the sighting wire in the cover. A prominent terrain feature on this line of sight is selected, the compass closed, and the landmark approached. When the scout reaches the landmark, the procedure is repeated.

Following an Azimuth at Night

It is necessary to prepare and set the compass before departing on a night movement because at night only the luminous parts of the compass can be seen. To prepare the compass for night use, the luminous parts must be fully charged by sunlight or artificial light, such as a flashlight. To set a compass—

1. Move the compass so that the desired azimuth on the dial is directly under the index line on the lower glass.

2. Rotate the upper movable glass so that the luminous line is directly above the north arrow of the dial.

3. Set the compass for marching at night on the specified azimuth.

—OR—

1. Face the general direction of movement.

2. Line up the north arrow and the luminous line on the bezel with the luminous sighting dots.

3. Hold the compass still with one hand and grip the knurled bezel ring with the other hand.

4. Turn the bezel ring the prescribed number of clicks in the proper direction, remembering that each click equals 3 degrees. For example, to set an azimuth of 21 degrees, the bezel ring would be turned seven clicks to the left.

5. Turn the whole compass until the north needle lines up with the luminous line. The compass is then set on the desired azimuth. The azimuth is the line formed by the two luminous sighting dots on the inside of the cover.

To march on a preset azimuth during night movement, open the compass and move it so the north arrow is directly below the luminous line (see fig. 2-4). Move

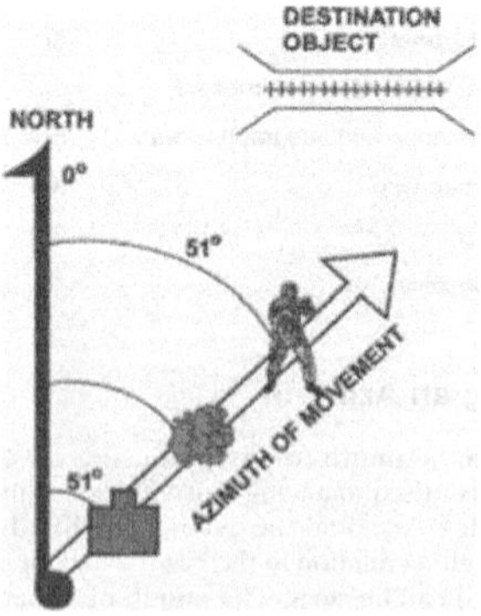

Figure 2-4. Following a Night Azimuth.

in the direction of the line formed by the two luminous sighting dots. It is necessary to refer to the compass more frequently at night than during the day. If stars are visible, find a prominent star along the azimuth of movement to use as a reference point. When the view of the sky is restricted by overcast conditions or vegetation, send a scout forward along the azimuth of movement to the limit of visibility. This scout is guided along the azimuth of movement by a stationary navigator. When the scout reaches the limit of visibility, the navigator moves to the scout's location. This process is repeated until the destination is reached.

A more rapid method for reaching the scout's destination is to equip the navigator with a compass. The navigator can set the compass as explained earlier and the scout proceeds providing security 180 degrees to the front on the specified azimuth, receiving right and left corrections from the navigator while both are on the move. The point scout must stay within visual range of the navigator. If available, a strip of white or luminous tape on the back of the point scout's helmet will assist.

Intersection

Intersection is the location of an unknown point by successively occupying at least two, preferably three known positions and sightings on the unknown point. It is used to locate features not depicted on the map or not readily identifiable. To determine an intersection, perform the following steps (see fig. 2-5):

1. Orient the map using the compass.

2. Locate and mark your position on the map.

3. Measure the magnetic azimuth to the unknown position; then convert to grid azimuth.

4. Draw a line on the map from your position on this grid azimuth.

5. Move to a second known position from which unknown point is visible. Locate this position on the map and again orient the map using the compass. The second unknown position should be a minimum of 30 degrees offset from the first position.

6. Repeat steps 4 and 5.

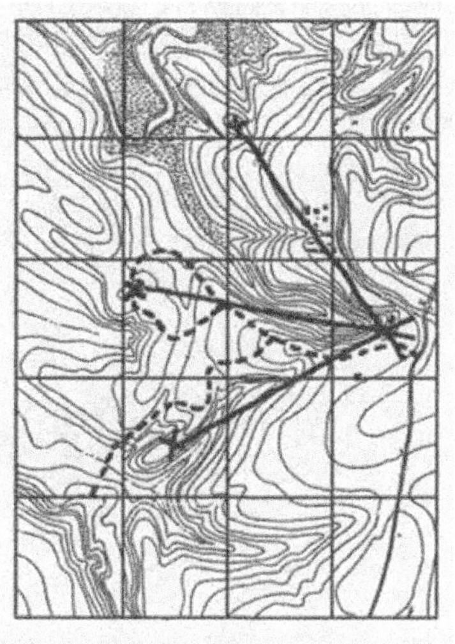

Figure 2-5. Intersection.

To check accuracy, move to a third position and repeat steps 1 thorough 4. Where the lines cross is the location of the unknown position. Using three lines, a triangle is sometimes formed—called the *triangle of error*—instead of an intersection. If the triangle is large, recheck your work to find the error. Do not assume that the position is at the center of the triangle.

Resection

Resection is the location of the user's unknown position by sighting on two or three known features that are identifiable on the map. To determine a resection, perform the following steps (see fig. 2-6):

1. Orient the map using the compass.

2. Locate two or three known positions on the ground and mark them on the map.

3. Measure the magnetic azimuth to a known position then convert to grid azimuth.

4. Change the grid azimuth to a back azimuth and draw a line on the map from the known position back toward the unknown position.

5. Repeat step 3 and step 4 to determine a second known position.

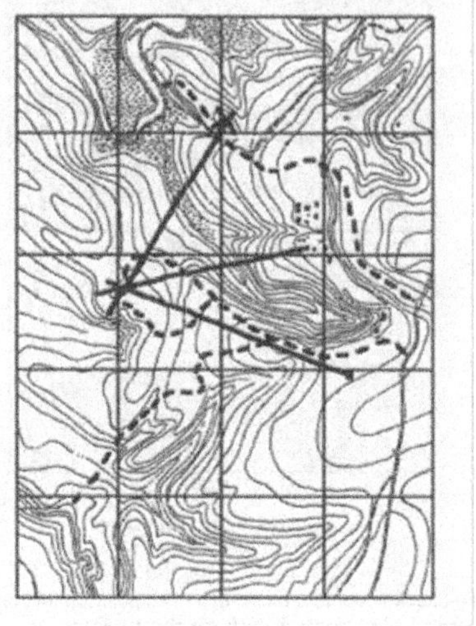

Figure 2-6. Resection.

To check accuracy, repeat the steps above for a third known position. The intersection of the lines is your location. Using three lines, a triangle of error may be formed. If the triangle is large, recheck.

2003. STARS, SUN, AND OTHER FEATURES

In rare cases when a scout is without a compass, the following examples are alternate means to determine direction. When using constellations to determine direction, identify your location's Temperate Zone. A Temperate Zone is the area between the tropics and the polar circles.

At night, the stars provide an excellent means of maintaining a line of march. In the North Temperate Zone (north of the equator), the Big Dipper constellation is one key to determining direction of true north. It is made up of seven fairly bright stars in the shape of a dipper with a long curved handle (see fig. 2-7). The two stars that form the side of the cup farthest from the handle, used as pointers, are situated in the direction of a bright star that is about five times the distance between the two stars of the dipper cup. This bright star is the North Star and is directly over the North Pole. The pointers always designate the North Star, which is the direction of true north.

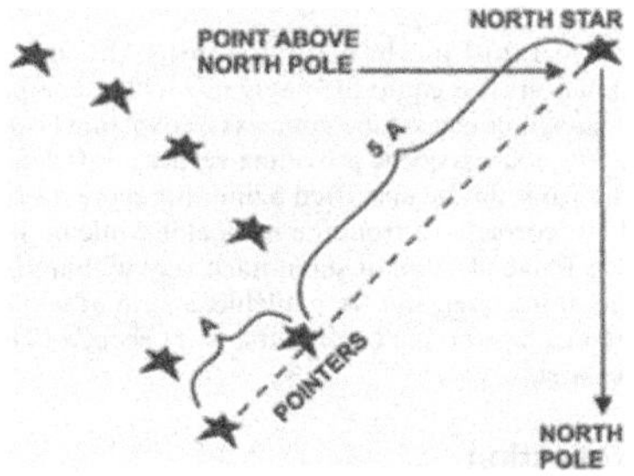

Figure 2-7. Locating the North Pole.

In the Southern Hemisphere, true south is determined in relation to the Southern Cross, a constellation composed of five stars. Two bright pointer stars in the vicinity of the Southern Cross serve as locators to help locate true south (see fig. 2-8). The outer four stars are

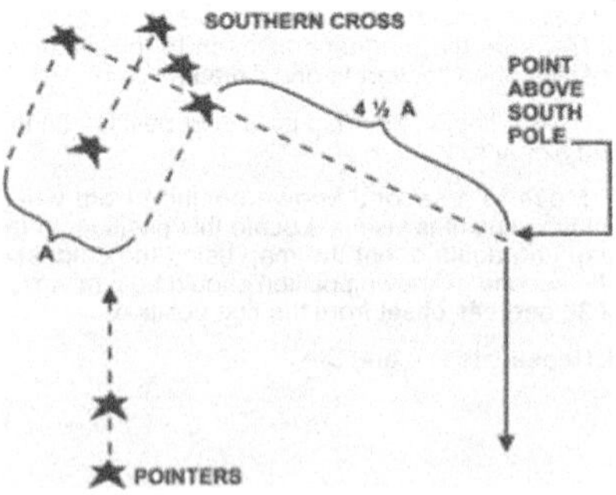

Figure 2-8. Locating the South Pole.

fairly bright and form a cross. This cross is imagined as the frame of a kite. A straight tail, four and one half times as long as the length of the kite itself, is put on the kite using finger widths for a measuring stick. The end of this tail will be close to a position directly over the South Pole. Usually, it will not be possible to see a star in the immediate vicinity, because there is no bright star visible directly above the South Pole.

During daylight hours, a watch and the sun can be used to determine direction within 8 degrees. In the North Temperate Zone, the watch is held horizontally, face up, and the hour hand pointed at the sun (see fig. 2-9). The north-south line and the direction of south can be found midway between the hour hand and the number 12, if the watch is set on standard time. If in daylight savings time, the direction of south is found midway between the hour hand and the number 1.

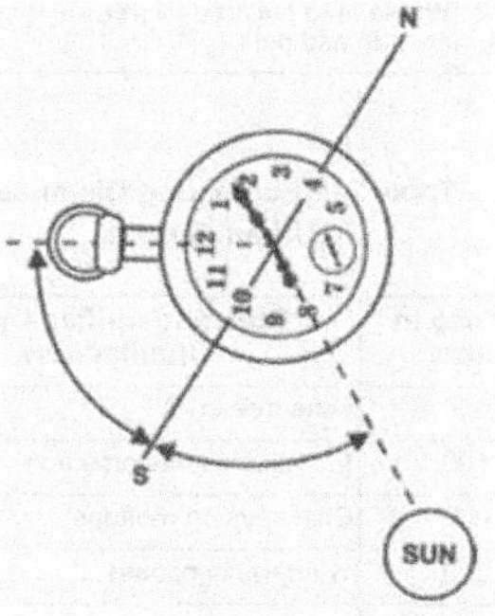

Figure 2-9. Determining Direction by Watch and Sun (North Temperate Zone).

In the South Temperate Zone, if the watch is set on standard time, the number 12 on the watch is pointed at the sun; if the watch is set on daylight savings time, the number 1 is pointed at the sun. North is midway between 12 (or 1) and the hour hand (see fig. 2-10).

When laying in a north-south line, if any doubt exists as to which end of the line is north, remember that the sun is in the east before noon and in the west in the afternoon.

In addition to the sun and stars, other methods a scout without a compass can use to determine direction include determining prevailing wind direction and

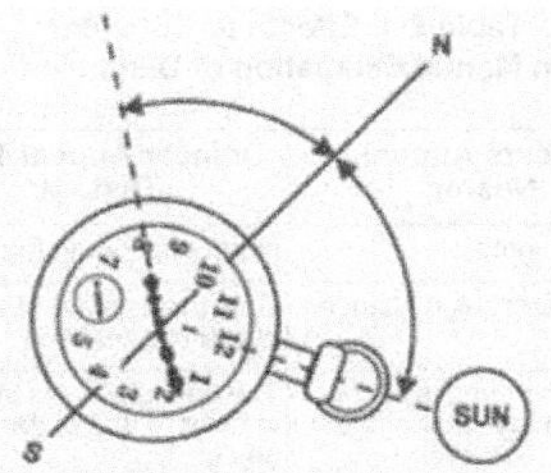

Figure 2-10. Determining Direction by Watch and Sun (South Temperate Zone).

using a mountain for orientation. By previous study of maps and photographs, a scout can keep informed of location and direction by using a distinctive edge of woods, a deep ravine or the direction of a stream's flow. A scout should constantly evaluate and memorize both the immediate terrain and general area for prominent features and landmarks.

2004. RANGE DETERMINATION

Range determination is the method of finding the distance between an observer and an enemy target or an object. By accurate range determination, the members of a given unit can set their sights correctly and place effective fire on enemy targets. The degree of accuracy is dependent on several factors, such as terrain relief, time available, and experience of the observer.

Mental Estimation

A mental distance estimate is made using a known unit of measure. Distance is estimated to the nearest 100 meters by determining the number of known units of measure between the observer's position and a target. For example, a football field, which is 100 yards, can be used as a known unit of measure for determining the distance between an observer's position and a target. For longer distances, progressive estimation may be necessary. To do this, the observer determines the number of units of measure to an intermediate point and doubles the value. The observer should consider the effects in table 2-1 in estimating distances.

Table 2-1. Effects to Consider in Mental Estimation of Distances.

Objects Appear Nearer	Objects Appear More Distant
In bright light.	In poor light or in fog.
In clear air at high altitude.	Only a small part of the object can be seen.
The background is in contrast with the color of the object.	The background is similar in color to that of the object.
The observer is looking down from a height.	The observer is looking over a depression, most of which is visible.
The observer is looking over a depression, most of which is hidden.	The observer is kneeling or sitting, especially on a hot day, when the ground is moist.
The observer is looking down a straight feature such as a road.	
The observer is looking over water, snow, or a uniform surface such as a cultivated field or desert.	

Estimating in Good Visibility

When visibility is good, distances can be estimated by using the appearance of tree trunks, branches, and foliage (as seen by the naked eye) in comparison with map data. Table 2-2 is a guide for wooded terrain. Table 2-3 is a guide for urban environments.

Estimating From a Terrain Study

The Marine should always use terrain/map analysis to assist in estimating distances. When the Marine is looking in a specific direction, the estimation of distance can be enhanced by studying the terrain and comparing it with the map. Particular emphasis should be given to color contrasts of terrain features seen along the observer-target line (OTL). For example, the distance across successive ridge lines or depressions in the distance may be identifiable by only slight changes of color to the eye. Different colors of grass might reveal a hidden terrain feature such as a stream.

Table 2-2. Estimating Distance in Wooded Terrain.

Distance in Meters	Tree Description
1,000	Trunk and main branches are visible. Foliage appears in cluster-like shape. Daylight may be seen through the foliage.
2,000	Trunk visible, main branches distinguishable, foliage appears as smooth surface. Outline of foliage of separate trees distinguishable.
3,000	Lower half of trunk visible. Branches blend with foliage. Foliage blends with adjoining trees.
4,000	Trunk and branches blend with foliage and appears as a continuous cluster, smooth in appearance. Movement of foliage due to wind cannot be deleted.
5,000 and beyond	Whole area covered by trees and appears smooth and dark.

Table 2-3. Estimating Distance in Urban Terrain.

Distance in Meters	Object Identified by the Unaided Eye
1,000	Lone tree trunk
1,500	Individuals and horsemen
3,000	Chimneys on rooftops
4,000	Windows in houses
4,000–5,000	Individual houses in populated area
8,000–9,000	Villages and individual houses
15,000–18,000	Large houses, towers, and steeples

Chapter 3. Enemy Activity

A commander often acts on information furnished by scouts. Therefore, scouts must aim at absolute accuracy in reporting enemy activity. This chapter discusses estimating enemy strengths, interpreting signs and tracks, and knowing the enemy.

3001. ESTIMATING ENEMY STRENGTHS

If troops cannot be counted, their strength may be estimated by: noting the length of time it takes various types of moving columns to pass given point, the area required of a unit in camp or bivouac, or the front on which they are deployed. When the ground is dry, infantry on the march raise a low, thick cloud of dust, and motor vehicles or mechanized units raise a thick, rapidly moving cloud. Additionally, through practice, a scout may gain information as to the strength and composition of enemy forces by listening to noises and observing lights, fires, and smoke. A scout gains valuable experience in estimating enemy strengths by observing friendly forces in camp, on the march, and deployed. The knowledge scouts gain during field exercises of the appearance and tactical dispositions of squads, platoons, companies, and larger units will be of great assistance in estimating the strength and composition of enemy units observed under various conditions.

3002. INTERPRETING SIGNS AND TRACKS

In addition to estimates made through direct observation, a scout may often be able to estimate size, composition, direction, rate of movement, condition, discipline, state of training, and morale of enemy forces through signs and tracks left behind.

Signs

The examination of vacated enemy positions provides valuable information. The size of a bivouac or defense area ordinarily indicates the number of enemy occupants. Clothing, ration containers, dumps, etc., further indicate the quantity of the departed enemy force. The condition of the bivouac area and amount of material abandoned give an indication of the enemy morale, training, and discipline. A well-policed area indicates good discipline. Rubbish, ration and smoking residue, and nonessential personal items of equipment adrift indicate a lower state of morale, training, and discipline. Stores and material left behind in good condition may indicate a hasty movement or withdrawal. Burned or destroyed materials indicate a deliberate, orderly withdrawal or movement. Letters, insignia, and other articles may reveal the identity of the enemy unit.

In the case of a moving enemy, the distance between periodic halts indicates the rate of march if enemy habits relative to marches and halts are known. Condition of the halt areas indicates the state of morale, training, and discipline.

The physical condition of enemy dead and wounded and their personal equipment and weapons are reported. The general condition and state of maintenance of destroyed or abandoned vehicles should also be reported.

Tracks

A track is a mark left on the ground by the passage of a person or object. Examination of tracks reveals information about the enemy.

Troops

A few tracks overlapping each other on both sides of a road or trail may indicate a patrol in staggered formation. A large number of tracks indicates troops in column formation. A large column will wear a dry road smooth and flat. In damp terrain, a freshly made track will have sharp edges; ordinarily, signs of moisture will disappear in about 15 minutes. A runner's toes are dug into the ground; a walker's footprint is fairly even.

Vehicles

The type of track indicates whether the vehicle is wheeled or tracked. A scout acquires the necessary experience to make the proper determination by observing vehicle tracks during training.

The direction of travel can be determined by the way tracks pass across ruts, by impressions on the edges of holes in the ground, how water is splashed from puddles, or by the way grass, twigs, and branches are bent; for example—

- A vehicle (wheeled or tracked) entering a rut pushes dirt into the rut and leaves an indentation on the exit side of the rut.
- A wheel going over holes in the ground leaves a deeper impression on the edge toward the direction of travel.
- The side of a puddle with the greater splash indicates the direction of travel of the vehicle.
- When traveling cross-country, the direction in which grass is bent and/or twigs, branches, and bushes are broken indicates direction of travel.

A general rate of speed can be estimated by the amount of water or mud splattered. A fast-moving vehicle will throw larger amounts of water or mud a greater distance to the front and sides than a slow-moving vehicle, and it leaves a deeper impression on the exit edges of holes. The faster the travel, the deeper the impression.

3003. KNOWING THE ENEMY

A scout should learn as much as possible about enemy psychology, habits, organization, and tactics. The more knowledge gained about the enemy, particularly the enemy's normal security measures, the better the scout's chances are to observe and obtain accurate information with minimum risk to the mission's success. Scouts gain much of this knowledge through experience, but they also gain a great deal of their preliminary information, particularly that pertaining to enemy organization and tactics, during training and may be updated by unit commanders and intelligence officers.

CHAPTER 4. DAYLIGHT SCOUTING

A scout must be able to operate in all types of terrain and under all conditions of visibility. He must be thoroughly familiar with the principles for using cover and concealment, camouflage, individual movement, and route selection, both to and from the objective.

4001. COVER AND CONCEALMENT

Cover is protection from the fire of hostile weapons. Concealment is protection from observation or surveillance from hostile air and ground observation, but not from hostile fire. Both cover and concealment are divided into two main categories: natural and artificial. Natural cover includes small hills, ditches, rocks or vegetation. Fighting holes, bunkers, and brick walls are examples of artificial cover. Some features, such as buildings, provide both cover and concealment. In deciding whether to seek cover or concealment, a scout must make the best choice to complete the mission (see fig. 4-1).

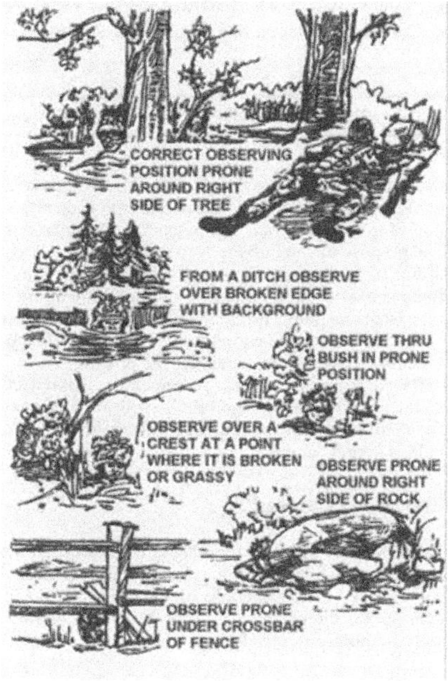

Figure 4-1. Correct Use of Cover.

Concealment Principles

Concealment principles are as follows:

- Remain motionless while observing. Anything in motion attracts the eye.
- Use all available concealment.
- Observe from the prone position (it offers a low silhouette and makes detection by the enemy difficult.
- Expose nothing that reflects light.
- Blend with the background because contrasting colors are noticeable.
- Remain in the shade because moving shadows attract attention.
- Distort or change the regular outline of objects. Most military objects have distinctive shapes that make obvious shadows and silhouettes.
- Avoid the skyline. Figures on the skyline can be seen from great distances and are easily identified by their outlines.

Concealment Techniques

Concealment techniques are as follows:

- When observing, the scout looks around an object's side (unless it is transparent) and prepares to fire, if necessary, around the side of or, if possible, through an object.
- Looking or firing over an object can make the scout an easily visible target for the enemy. If the scout must fire over the top of concealment or cover, the outline of the head or helmet should be broken or distorted.
- Upon the approach of an airplane, the scout takes a prone position, turns face-down, and remains motionless. If surprised by an airplane, the scout remains in place and does not look up.

- The scout covers exposed body parts such as the face, back of the neck, and hands with grease paint, mud or other materials to reduce sun reflection.
- Camouflage for equipment can be improvised from garnishing or sandbags to prevent sun reflection.
- In snowy terrain, white overgarments are worn.
- The helmet cover outline should be distorted.

4002. CAMOUFLAGE

Camouflage is the use of concealment and disguise to minimize the possibility of detection and/or identification of troops, material, equipment, and installations. The purpose of camouflage is to provide concealment of military objects from enemy observation. Camouflage is also used to conceal an object by making it look like something else. A scout's mission usually requires individual and equipment camouflage. If natural camouflage is not adequate, the position is camouflaged. In using camouflage, remember that objects are identified by their form (outline), shadow, texture, and color. The principal purpose of camouflage in the field is to prevent direct observation and recognition.

Individual Camouflage

Successful individual camouflage involves the ability to recognize and take advantage of all forms of natural and artificial concealment available (vegetation, soil, debris, etc.) and knowledge of the proper use of artificial camouflage materials.

Aids to Individual Camouflage

A scout must recognize the terrain's dominant color and pattern and must change the appearance of clothing and equipment accordingly in order to blend and not contrast with the terrain (see fig. 4-2).

The helmet is camouflaged by breaking up its shape, smooth surface, and shadow. Use of a helmet cover works best. In the absence of a helmet cover, mud can be irregularly blotched on the helmet to disguise its form and dull the surface. A helmet cover may be improvised from irregularly colored cloth or burlap to blend with the background. Foliage can be draped to prevent the visor of the helmet from casting a dark shadow across the face. Foliage should not stick up like plumes because any head movement will give away the position.

A small, thin bush in the shadow of a large bush makes a good observation point. Lone trees, rocks, fence corners, and outstanding landmarks are easily picked up by the enemy as obvious observation posts.

If camouflage clothing is not available, other available clothing can be attached in irregular splotches of appropriate colors.

Exposed skin reflects light and attracts the enemy's attention. Even very dark skin will reflect light because of its natural oil. The buddy system is recom-mended when applying camouflage. Standard Marine Corps issue camouflage face paint sticks are two toned:

- Loam and light green for light-skinned troops, in all but snow regions.
- Sand and light green for dark-skinned troops.
- Loam and white for troops in snow-covered terrain.

Shiny areas (forehead, cheekbones, nose, and chin) are painted with a dark color. Shadow areas (around the eyes, under the nose, and under the chin) are painted

Figure 4-2. Avoid Contrasting Backgrounds.

with a light color. Skin that is exposed on the back of the neck and hands is painted with a two-color combination in an irregular pattern (see fig. 4-3).

When standard issue face paint sticks are not available, burnt cork, charcoal or lamp black can be used to tone down exposed areas of skin.

Mud is used only in an emergency because it changes color as it dries and may peel off, leaving the skin exposed. Since mud may contain harmful bacteria, mud should be washed off as soon as possible.

Any equipment that reflects light should be covered with a nonreflective material that aids in the concealment of the weapon (for example, black electrical tape or mud). The straight line of the rifle or other infantry weapons may be very conspicuous to an enemy observer. The barrel and hand guard should be wrapped with strips of contrasting colored cloth or

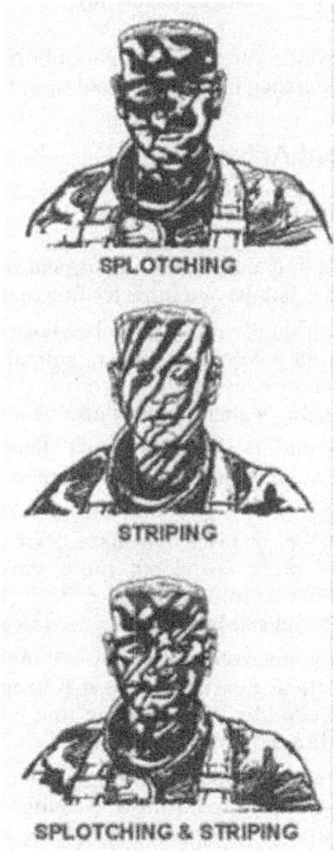

Figure 4-3. Face Camouflage.

tape to break the regular outline. Mud or dirt dulls the reflecting surface of the stock, barrel, and bayonet where coloring has been worn. Lamp black may also be used on metal parts. The function of the weapon must not impaired.

If time, material, and surroundings permit, a ghillie suit should be constructed. (Refer to MCWP 3-15.3, *Scout Sniping*.)

Aids to Camouflage a Position

To successfully camouflage a position, the scout must remember to—

- Camouflage the position as soon as it is occupied.
- Avoid using too much material for camouflage. Even though natural materials are used, too much may make the object and its shadow stand out from its surroundings, thus attracting the attention of a hostile observer.
- Inspect completed camouflage work from the enemy's point of view to check effectiveness.

Continuous Camouflage

Camouflage around and on the scout's position must be maintained in a fresh condition as wilted and dead foliage can give the position away. If the mission dictates that the position should be occupied for longer periods, wilted foliage should be replaced during periods of reduced visibility.

4003. INDIVIDUAL MOVEMENT

Principles

The principles of individual movement are as follows:

- Scouts move from one concealed position to another. When not changing positions, they remain motionless.
- The scout's head is lifted slowly but steadily, without abrupt movements, to search for a new position.
- Scouts select the next stopping place before moving and ensures it is not contained by the enemy.
- Scouts change position on the run: springs up, runs with the body bent low, zigzags, quickly drops to the ground slightly to the right or left of the objective, then rolls or crawls to the desired position.

(Remember the phrase, "I'm up—he sees me—I'm down.")

Rushing

When starting from the prone position—

- Raise the head slowly and steadily and select a new position.
- Lower the head slowly, draw arms inward, cock right leg forward, and prepare to rush.
- Use one movement to raise the body by straightening both arms.
- Spring to your feet, stepping off with the left foot.
- Bend forward as low as possible when running. Never advance directly to the next position; always zigzag.

When hitting the deck—

- Stop.
- Plant both feet in place.
- Drop quickly to the knees and slide the hand to the heel of the rifle.
- Fall forward, breaking your fall with the butt of the rifle. (To confuse the enemy, roll over after hitting the deck and roll into firing position with feet, knees, and stomach flat on the ground.)
- Keep head down if you do not intend to fire.

When rolling over—

- Hit the deck and assume the prone position.
- Bring the rifle in close to the body, placing the rifle butt in the crotch.
- Roll over swiftly to confuse any enemy observers as to final intended location. Never reappear at the same place you went down.

Low Crawl

The low crawl is used when—

- Cover and concealment are scarce.
- The enemy has good observation over the area in which the scout is moving.
- Speed is not essential.

To perform the low crawl, keep the body as flat as possible against the ground. Grasp the rifle sling at the upper sling swivel. Let the balance of the rifle rest on the forearm and let the butt of the rifle drag on the ground. Keep the muzzle off the ground.

To start forward, push arms forward and pull right leg forward. To move forward, pull with arms and push with right leg. Change the pushing leg frequently to avoid fatigue.

High Crawl

The high crawl is used when—

- Cover and/or concealment are available.
- Poor visibility reduces enemy observation
- Greater speed of movement is required.

To perform the high crawl, keep body off the ground. Rest weight on forearms and lower legs. Cradle rifle in arms, keeping the muzzle off the ground. Keep knees well behind the buttocks to stay low.

Move forward, alternately advancing right forearm and left knee; then left forearm and right knee.

Movement Aids

Aids to movement include—

- Carrying only necessities. Additional weight causes premature fatigue and impedes free movement.
- Not disturbing birds or animals whose flight would betray your presence. If birds or animals are alerted, remain motionless under cover for a few minutes, as the enemy's attention may also be attracted.
- Moving during an incident that diverts attention, such as an airplane flight, a distant disturbance or sudden bursts of fire.
- Fog, smoke, or even light haze offer concealment for movement; however, the enemy may have thermoimagery and night vision devices. Therefore, darkness and smoke cannot be used as easily.
- Following a stream or road by staying as far away from them as possible while still keeping them in sight. Keep close to the dune line when moving along a beach.
- When moving through tall grass or similar growth, move when the wind blows, changing direction frequently. A straight route will be more readily noticed.
- Whenever possible, avoid areas of soft ground so as not to leave tracks.

- When crossing a road or water obstacle, choose crossing sites where the enemy's observation is restricted (an area in shadows or near a bend) and cross rapidly.

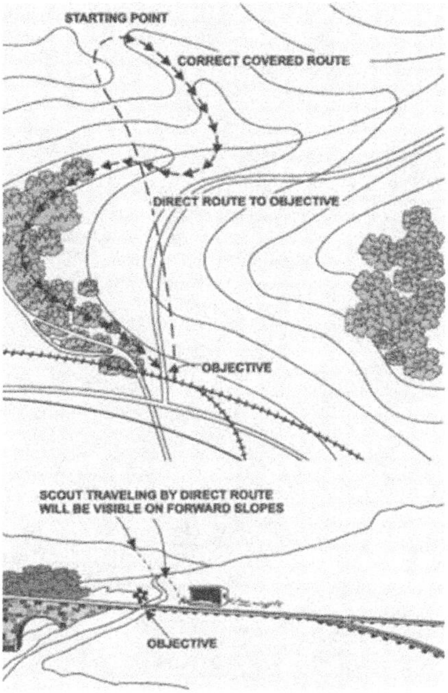

Figure 4-4. Choosing a Concealed Route of Advance from a Map.

4004. ROUTE SELECTION

Prior to Movement

A scout and the immediate commander conduct a map reconnaissance before starting on a mission. This assists them in selecting the route according to available cover and concealment and any indicated enemy activity.

Prior to and during the course of the mission, move to an observation point to visually reconnoiter the terrain for movement and select the tentative route. It may be necessary to make wide detours around open spaces or those containing enemy patrols or other enemy activity.

Carefully study the country to be traversed and pay close attention to the general features, streams, ridges to be crossed, and their relation to the general direction to be taken (see fig. 4-4).

Make notes of terrain features and landmarks along the proposed route and rely on notes for guidance (see fig. 4-5 on page 4-6). Additionally, determine the compass direction and readings for each change of direction at the start. Finally, learn the location of unit boundaries and observation/listening posts as well as general location of other friendly or scouting parties. Be sure to avoid man-made and natural obstacles as they will slow progress and overall success of the mission. If possible, use the local populous as a source of intelligence. When returning to friendly lines, avoid using the same route.

En Route

En route, the actual advance will be a series of movements from one observation point to the next. The distance and route will depend on cover and terrain. Assess the cover, terrain, and any enemy or civilian activity to determine whether or not to modify the approach or return routes. Unless the mission requires it, avoid danger areas (for example, houses, villages, potential assembly or bivouac areas, roads, and streams) that may give away your position by being observed by the enemy. When required to reconnoiter danger areas, choose a covered approach and return, and make entry or passage as quietly and quickly as possible. If part of a larger effort, the approach and return should be covered by observation and fires of the other members of the scouting party or patrol.

Stream Crossings

When the crossing does not appear to be held by the enemy, advance upon it rapidly. If there are two or more scouts, one crosses while the other(s) provide protection. Note the length, width, depth, and approaches to a crossing. Observe the condition of the road or trail that crosses the stream, and report on the suitability of the crossing for use by tracked and wheeled vehicles. If the crossing is under observation by enemy, seek another crossing site or dash across to avoid detection.

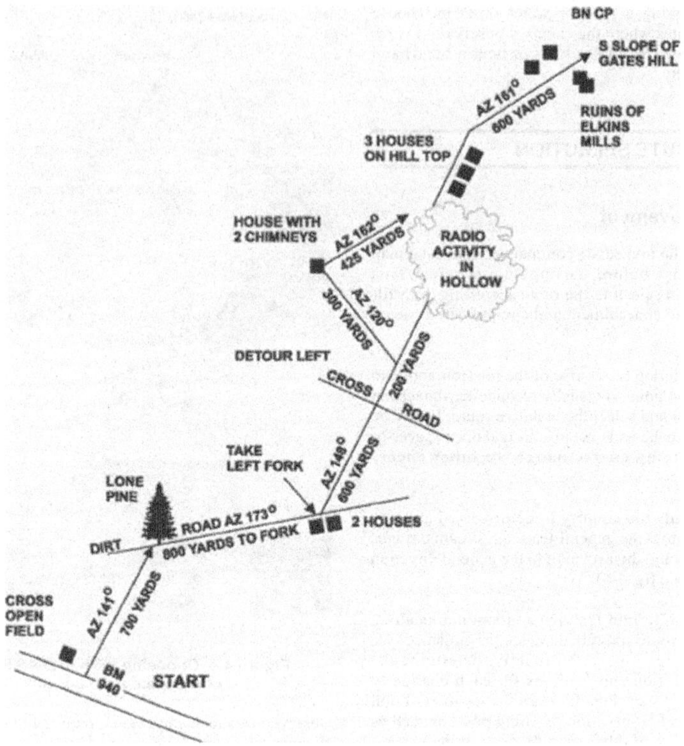

Figure 4-5. Proposed Route Sketch.

CHAPTER 5. NIGHT SCOUTING

Night scouting presents many of the same problems encountered in day operations-such as cover, concealment, movement, and camouflage-as well as additional considerations. Knowledge of human eye construction and operation will enable maximum advantage under night conditions or poor visibility.

5001. NIGHT VISION

Certain parts of the eye correspond to parts of a simple camera (see fig. 5-1). The lens focuses light entering the eye similar to a camera lens. The iris (colored part of eye) corresponds to the diaphragm of a camera, opening and closing to regulate the amount of light entering the eye through the pupil. The retina corresponds to camera film. Light rays strike the retina, form an image, and cause an impression to be transmitted to the brain through the optic nerve. In a camera, the image is formed and fixed on film.

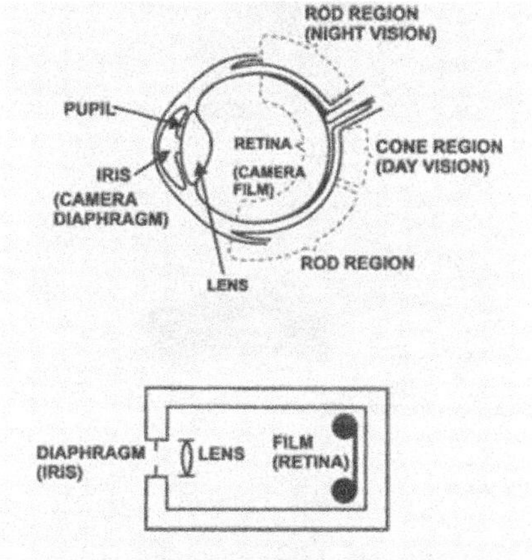

Figure 5-1. The Eye is Like a Camera.

The retina is composed of cone cells and rod cells, so-called because of their shapes. Cone cells distinguish color, shape, and sharp contrast. Because they are activated by light conditions, they are blind during periods of low illumination. Rod cells produce a chemical substance called visual purple that makes them active in darkness, low illumination or night conditions. Rod vision distinguishes black, white, shades of gray, and general outlines.

Principles

To effectively "see" at night, the principles of night vision dark adaptation, off-center vision, and scanning are applied.

Dark Adaptation

Allowing the eyes to become accustomed to low levels of illumination is called dark adaptation. It takes the rod cells about 30 minutes to produce enough visual purple to activate them and enable the eye to distinguish objects in dim light. This may also be accomplished by staying in a red-lighted area, or by wearing red goggles for 20 minutes, followed by 10 minutes in darkness (which allows the pupils to open wide). This method saves valuable time by allowing Marines to be in a lighted area to receive orders, check equipment, or perform some other function before moving into darkness.

Off-Center Vision

The technique of focusing on an object without looking directly at it is called off-center vision. When looking directly at an object, the image is formed on the cone region, which is not sensitive at night (see fig. 5-2 on page 5-2). When looking slightly to the left, right, above or below an object, the image is formed on the area of the retina containing rod cells, which are sensitive in darkness. The most sensitive area varies in individuals, but is usually found by looking 6 to 10 degrees away from an object; in effect, out of the corner of the eye (see fig. 5-3 on page 5-2).

Scanning

Off-center vision used to observe an area or an object is called scanning. When using rod vision, the visual purple in the rod cells bleaches or blacks out in 4 to 10 seconds and the object observed disappears. As the visual purple in the rod cells in one area bleaches out, the eyes must slightly shift to use fresh rod cells. Eyes

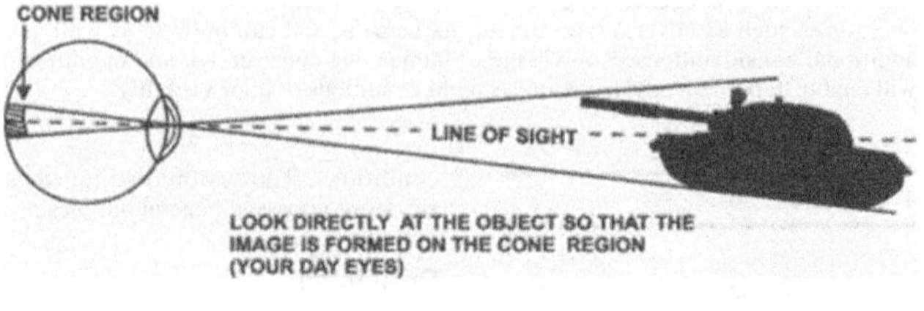

Figure 5-2. Day Vision.

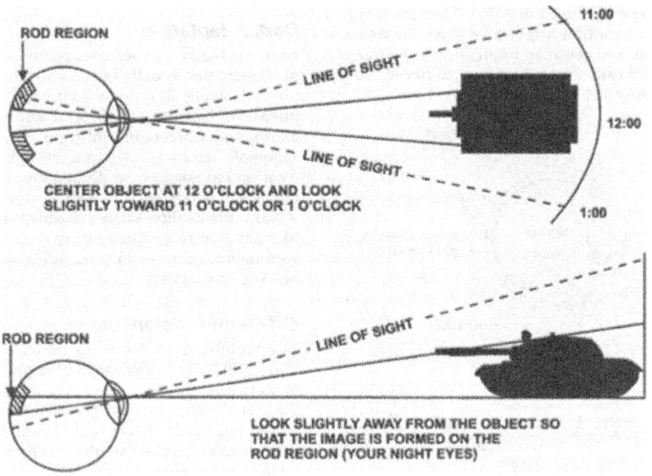

Figure 5-3. Night Vision.

should be moved in short, abrupt, irregular movements over and around the target (see fig. 5-4).

Preserving Night Vision

Night vision is quickly destroyed if bright light is allowed to enter the eye. When entering a lighted area or when observing in a temporarily lighted area (illumination, flares), one eye should be closed and covered to preserve its night vision. When the light goes off, fades or the lighted area is exited, the night vision retained by the protected eye enables it to see until the other eye adapts to the darkness. Red light helps preserve night vision, but like white light, it can be observed at long distances.

Factors that decrease night visual acuity include fatigue, lack of oxygen, long exposure to sunlight, alcohol, nicotine (within the past 48 hours), and age. When night vision has been attained, straining will not improve effectiveness; however, practice identifying objects at night will improve perception.

5002. APPEARANCE OF OBJECTS

Darkness not only makes objects difficult to see but also changes their appearance, distorts size, and blots out details. A tree visible against the night sky appears smaller than in the daytime because the twigs at the end of branches cannot be seen. A scout must train to identify objects by block outlines at night and cannot rely on details visible in daylight. Binoculars enlarge objects or parts of objects otherwise too small to be seen and help identify objects already spotted. Night observation devices increase night visibility and should be used whenever possible.

5003. SOUNDS

At night, sounds become very important. By listening, a scout gains information about the enemy and by exercising care, keeps information from the enemy. A scout stops frequently to listen. Scouts must listen for long periods in perfect silence. Hearing is amplified with the mouth open; removing the helmet will reduce sound distortion. Sounds are transmitted a greater distance in wet weather and at night than in dry weather and in the daytime. By holding the ear close to the ground sounds of people walking and vehicles moving can be heard. Sound travels approximately 370 meters a second. When a flash from a fired weapon is observed, the range to the weapons can be easily estimated by counting the time interval between the flash and hearing the report. For example, counting to three (one thousand one, one thousand two, one thousand three), indicates the distance is 1,110 meters. The cadence is determined by actual practice at known ranges.

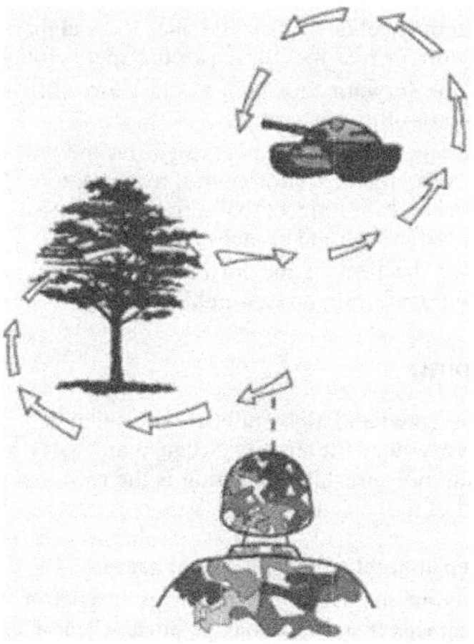

Figure 5-4. Scanning

5004. SMELLS AND TOUCH

A scout's sense of smell can warn of enemy fires, cooking, motor parks, gasoline and diesel engines, and bodies of water. A scout must feel and recognize objects in the dark, and adjust and operate equipment quietly by sense of touch.

5005. CLOTHING AND WEAPONS

All loose clothing must be secured (string or tape can be used) to prevent snagging on barbed wire, brambles, and brush. Helmet covers are worn to muffle sounds made by low branches.

The belt buckle should be turned around to the side in order to move in a prone position without scraping the buckle against stones or hard surfaces. Identification tags can be taped together to prevent rattling. Hands, face, and neck can be blackened so skin does not reflect light or appear as white spots in the darkness. (Refer to para. 4002.)

Scouts are normally armed with rifles. Rifle slings should be taped to prevent rattling. All weapons parts should be checked for glare elimination measures.

5006. CONCEALMENT

Although total darkness provides concealment, scouts must observe the same principles of concealment during moonlight conditions as in the daytime. Scouts should assume enemy employment of night observation devices and observe the principles of night movement such that presence will not be disclosed by noise when close to the enemy.

5007. AIDS TO NIGHT SCOUTING

Aids to night scouting include the following:

- Carry out scouting missions close to or within hostile positions on dark or rainy nights.
- Stifle a threatening sneeze by pressing fingers upward against the nostrils.
- Stifle a threatening cough by applying slight pressure with the finger on the Adam's apple.
- Stop a ringing sound that interferes with hearing by yawning.
- Speak softly rather than whispering if voice communication is necessary.
- Move boldly and rapidly when taking advantage of any sound-such as shelling, rustling wind or distant firing-to push forward if firing is taking place.
- Avoid shell craters and depressions in damp weather conditions if the enemy has employed chemical munitions.
- Move the eyes constantly; concentrating on one object too long will strain them.
- Take notice of the enemy's use of flares. When the enemy employs flares, few enemy patrols are apt to be out; when flares are not employed, the enemy's patrols are likely to be numerous.
- Drop to a prone position upon hearing a flare being fired and before it illuminates. Remain motionless while it is burning. If open or moving when a flare bursts in the air, freeze or drop quickly in the split second after the flare illuminates while the enemy is blinded. You are an easy target for the enemy if the flare bursts in the air or on the ground behind you. Never look at a flare. If you activate a trip flare, drop to the ground and crawl away from the illuminated area.
- Consider all patrols or individuals encountered as hostile until proven friendly. If encountering someone, crouch low, silhouetting the approaching person against the sky. At the same time, make yourself an indistinct target in case the person encountered is an enemy.
- Return fire only to avoid capture if fired on when close to enemy positions.

5008. AIDS TO NIGHT MOVEMENT

- Aids to night movement include the following:
- Move silently.
- Advance in stealthy legs. Each leg should follow some terrain feature that serves as a guide. When there are no terrain features to serve as guides, move in a straight or nearly straight line from one defined point to another, or maintain direction by using a compass.
- Avoid running, except in an emergency.
- Take advantage of sounds that may distract the enemy.
- Fall silently without making an outcry.

Walking

When walking at night—

- Place the heel down first. Balance the weight of the body on the rear foot until a secure spot is found.
- Lift the forward foot high to clear any stiff grass, brush, or other obstruction.
- Continue to balance body weight on the rear foot, lower the forward foot gently, toe first, to explore the ground for objects that might make noise. Step over fallen logs and branches, not on them.
- Lower the heel of the forward foot slowly; gradually transferring body weight to that foot.

Creeping

The low crawl and high crawl are not suitable at night when very near the enemy because an easily heard shuffling noise results. Creeping is the recommended method of movement:

- Creep at night on the hands and knees.
- Use your hands to feel for twigs, leaves or other substances that might make a noise. Clear a spot to place your knee. Keeping your hand at that spot,

bring your knee forward until it meets your hand. Then place your knee on the ground and repeat the action with the other hand and knee.

- Lay the rifle on the ground at your side and clear an area for it. Lift the rifle up and move it forward. Movement is slow and tedious, since it must be done silently.

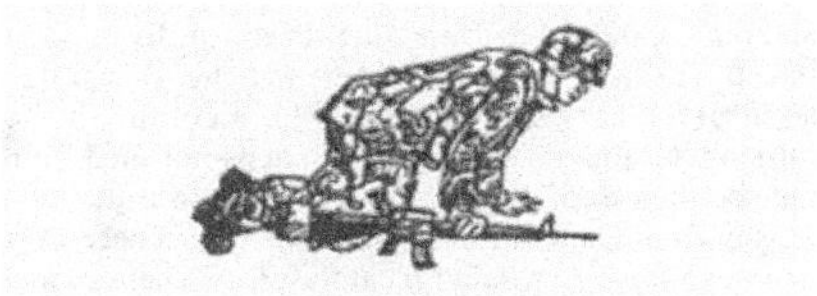

Hitting the Deck at Night (Right-Handed Shooter)

To safely hit the deck at night from the standing position—

- Advance your left leg, place the butt of the rifle in your right armpit with the hand remaining on the pistol grip, and grasp it with the right hand at the balance.

- Quietly drop down on the right knee and left hand.

- Move the left leg carefully to the rear, and then move the right leg to the rear.

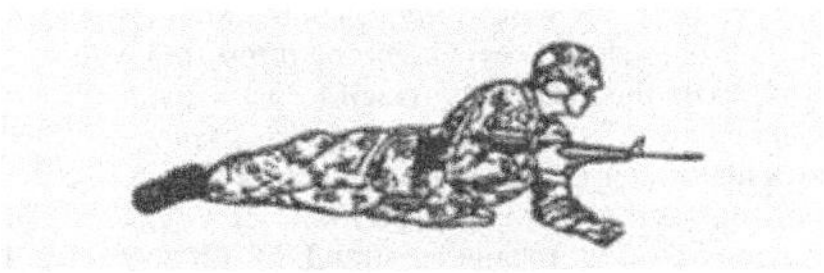

- Lie flat on the ground, or take up a firing position if necessary.

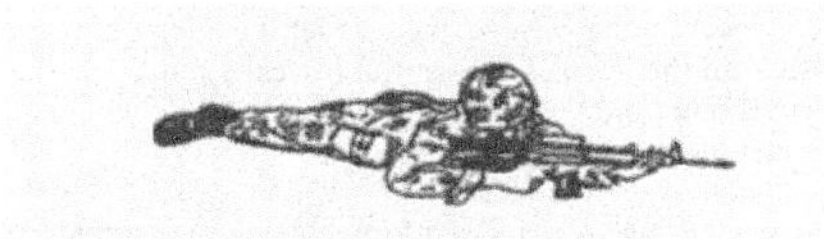

Wire Obstacles

A mission often requires a scout to pass through and work behind enemy positions. To accomplish this, the scout must be able to quietly pass through enemy wire obstacles and cross trenches. Cutting a gap in wire is time-consuming. If possible, walk over the low bands of enemy wire and crawl under the high bands (see fig. 5-5). Avoid movement along wire barriers, as enemy

Figure 5-5. Crossing Wire Silently at Night.

covering fires are generally planned parallel to them to take advantage of canalization and enfilade fire.

To step over low wire at night, crouch low to view the strands against the sky. Grasp the top strand with one hand; with the other hand, reach forward and feel for a clear spot for foot placement without stepping on other strands or any object apt to make a noise. Raise the body up, still grasping the top strand of wire. To avoid catching the foot in another strand, lift the foot up and over, passing it close to the hand grasping the wire.

If a high wire obstacle is encountered at night and wire cutters are not available, pass under the wire with your back on the ground. Grasp the lowest strands in your hands and hold them clear of the body while you slide under them.

When cutting wire and working solo, cut a wire near a post (see fig. 5-6), then dispose of all but one loose end. Grasp the wire close to a post and cut between your hand and the post, muffling the sound and keeping the loose wire in your grasp. When cutting wire in tandem, one firmly holds the wire with the hands positioned close to the cutters, in order to muffle the sound and prevent the loose ends from flying back, while the other one cuts. In both instances, the loose ends of the wire are bent back to form a passage.

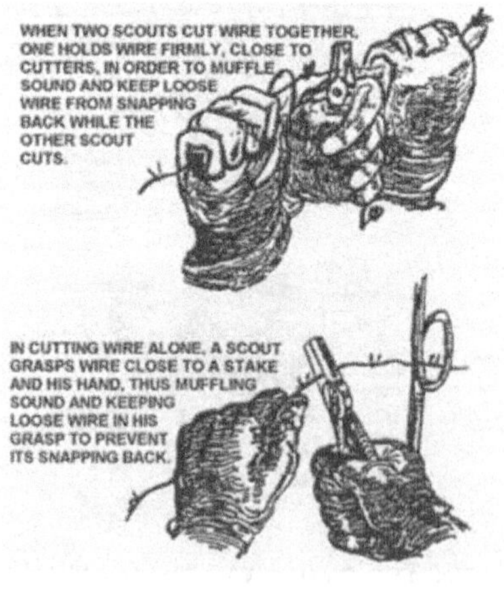

Figure 5-6. Cutting Wire Silently at Night.

Wrap a sandbag around the wire cutters and wire to deaden the sound.

Do not cut a complete gap in the wire; cut only the bottom wire(s). Leave the top wire(s) intact to lessen the chance of discovery by the enemy.

Crossing Trenches

Before approaching a trench, wait outside the trench for awhile and listen. Do not enter or cross a trench near its junction with a communication trench. Crawl silently up to the edge of the trench and look into it. Remove all loose dirt and rocks from the edge. If it is a narrow trench, spring up and jump across, sinking quietly to the ground on the other side and remaining there a moment to listen before proceeding. If the trench is wide, climb silently and slowly down into it and out the other side, using the revetment for support (see fig. 5-7). Do not enter enemy trenches unless it is absolutely necessary in order to accomplish the mission. Ordinarily, work is better accomplished from outside the trench. Sentries usually pay more attention to sounds in front of them; therefore, if it is necessary to enter a trench, cross it first at the place where enemy observation is restricted, then approach from the rear.

5009. LOCATING AND PLOTTING THE ENEMY AT NIGHT

For night work, a scout must understand the use of a lensatic compass. (Refer to para. 2002.) Using the lensatic compass, the scout can guide the platoon into position, locate adjoining elements of the command, keep direction when on patrol, determine the location of gaps in the enemy wire and the position of enemy out guards.

Locating Gaps in Enemy Wire

When searching for gaps in enemy wire, at least two lensatic compasses are needed: one to register the gap in the wire and the other for navigation. When a gap in enemy wire is located, lie outside the gap, keeping a distance of 10 meters from the barbed wire. Sight with the lensatic compass on a prominent point on the skyline in line with the gap. In selecting the prominent point in the skyline, pick one that appears on the map (i.e., hill mass, house, road junction). If the only prominent point available is one not identifiable on a map (i.e., a tree, destroyed vehicle, enemy position),

Figure 5-7. Crossing Trenches Silently at Night.

use it. The exact location of this point can be fixed the next day by visual reconnaissance of the area from an observation point. When the needle rests—

- Clamp it in place by lowering the eyepiece to the closed position.
- Rotate the movable bezel ring until the luminous line is directly over the north end of the needle. The azimuth of the gap from the prominent point is now registered.

The compass is carried back without further adjustment. The azimuth setting can be recorded later on a map (see fig. 5-8).

Locating Enemy Out Guards

At night, locate enemy guards by their sounds and failure to maintain light discipline. As sounds of the enemy are heard, and/or observations of the enemy made, shoot an azimuth with one compass. When the needle comes to rest, clamp the compass by lowering the eyepiece to the closed position.

Figure 5-8. Locating a Gap in Enemy Wire at Night.

Note the time and nature of each sound, the estimated distance, and which compass was used to fix the location. Plot this data on a map then wait until the debrief to turn in the notes and compasses. See figure 5-9.

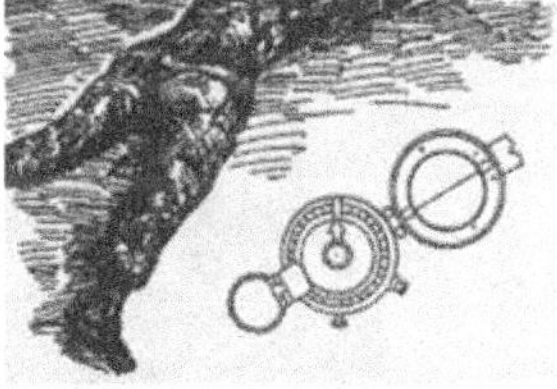

Figure 5-9. Locating Enemy Outguards at Night.

50010. ROUTES OF MOVEMENT

Prior to beginning a night missions, a scout studies the ground in detail from an observation point, air photos, and a map during daylight. The route of advance should be below the skyline. Avoid becoming a silhouette (see fig. 5-10).

Unless the moon is bright, avoid, if possible, passing through woods, ditches, ravines, and brush, because noises of movement may lead to discovery. If the enemy is known to have night observation device capability, avoiding these kinds of terrain may not be possible. To avoid enemy ambushes, return by a different route from the advance and change routes on successive nights.

Figure 5-10. Difference Between Correct Routes by Day or Night.

CHAPTER 6. OBSERVING AND REPORTING

When sent out on a mission, a scout's duties are to observe and report. the first five chapters covered scout protection measures and movement in enemy territory. This chapter discusses performing the assigned mission and reporting the mission after its completion.

6001. OBSERVATION POSTS

Positions

The following guidelines apply to observation posts. See figure 6-1.

- When selecting observation posts, scouts should chose the least prominent position.
- Scouts may occupy one or more observation posts.
- An observation post should not be manned for more than 24 hours.
- A selected observation post should be observed for 10 to 15 minutes to ensure it is not occupied.
- Scouts move to the chosen observation post by a concealed route.
- If the post is located on a hill, crawl to a position where the skyline is broken.
- If a tree is used, the position should have a background so as not to be silhouetted against the sky while climbing or observing.
- When leaving the observation post, a different route from that of the approach should be used.
- If a radio is used, its antenna should be located to provide clear communication to the controlling commander but masked from enemy observation and direction-finding equipment. Upon departure, scouts should remove the antennae from the observation post so as not to give away the position.

Observing

Using all senses available, be particularly alert for movement, objects, sounds, and smells inappropriate to the surroundings. While observing, avoid all unnecessary movement. If observing from a building, keep back from doors and windows.

In daylight, look first at the ground nearest you. Begin observing close to your post and search a narrow strip 50 meters or less deep, going from right to left parallel to your front; then search from left to right a second and similar strip farther away but overlapping the first. Continue to observe until the entire field of view has been searched (see fig. 6-2 on page 6-2).

At night, use a night observation device. If one is not available, search the horizon with short, jerky movements, and short pauses. Look a little to one side of an object and then to the other. Lower the head close to the ground to view the object more clearly. Use low-powered field glasses to increase sight range.

Figure 6-1. Method of Approaching an Observation Post.

Figure 6-2. Method of Searching Ground.

6002. REPORTING

It is imperative that the scout accurately and completely report who, where, when, and what was observed to the proper authority upon mission completion. In most cases, a scout will not be equipped with a radio.

Verbal Reports

Verbal reports should be made when writing is impractical, when the information is not complicated or when the enemy is likely to intercept a messenger carrying a written message.

Written Messages

Written messages, preferred to verbal reports, are recorded in message book blanks issued for that purpose (see fig. 6-3) and delivered to a higher authority as soon as possible. The NATO spot report (SPOTREP) should be reviewed as the written message is a NATO format.

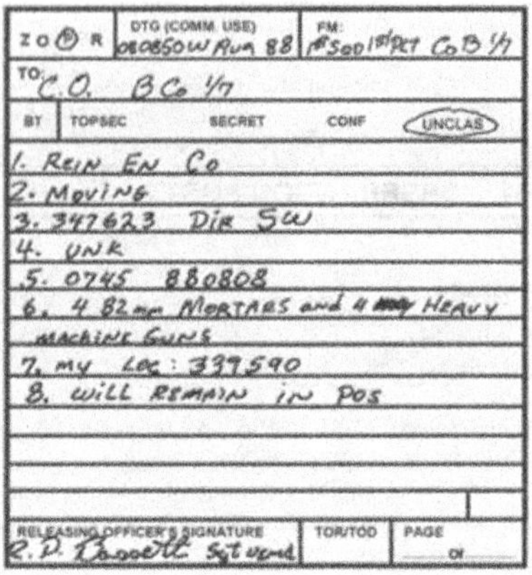

Figure 6-3. Sample Field Message.

The message body is brief, accurate, and clear; facts and opinions are distinguished. If secondhand information is reported, its source is included. Reports include all information of value, first about the enemy, and then about the originator. Information about the enemy should cover—

- Size and/or strength.
- Actions or activity.
- Location and direction of movement.
- Unit identification. (The designation of the enemy unit may be derived from unit markings, uniforms worn or through prisoner interrogation.)
- Time of observation.
- Equipment and weapons.

Messages are printed in block letters. Individual items of information are numbered and separated into paragraphs. If doubt exists as to message receipt by the commander, a summary of its contents is included in the next message. Information about the originator or writer should cover—

- Location at the time of enemy observation (reference to an important terrain feature, by map coordinates, by the back azimuth from each of two definitely located points, or the back azimuth and distance from one known point).
- Intentions. (Remain in position? Continue on the mission? Take other action?)

The message is carefully reread and if possible read by another person to ensure understandability. If a messenger is used, the messenger must read and understand the message in order to answer any questions the commander might have.

Sketch

Information difficult to describe may be given accurately on a simple sketch. The sketch may give all the necessary information or it may be used to supplement a written message. A military sketch is generally one of two types: simple or panoramic. The simple sketch is easily made and read.

Figure 6-4 shows a simple sketch that has been included in the message itself. The sketch may be on a separate sheet of paper, but all of the necessary information must be contained in either the sketch, the message or both.

The panoramic sketch is a picture of the terrain's elevation in perspective, as seen from one point of observation. Although a panoramic sketch is not difficult to create, skill and training are necessary to enhance usefulness. Figure 6-5 on page 6-4 illustrates how to make a panoramic sketch.

Overlay

The same information sent back to higher headquarters on the sketch may be sent on an overlay, if the sender and the person to whom the message is to be sent have copies of the same map. Figure 6-6 on page 6-6 illustrates a simple overlay. The overlay is drawn on transparent paper as follows:

- Orient the map and place it on a hard, flat surface.
- Place the transparent paper over the part of the map of the object or information to be transmitted and hold the paper in this position.
- Orient the overlay to the map by tracing in the intersecting grid lines at two opposite corners of the overlay. Write the correct number designation on the overlay. The cross made by the intersection is called a *tick mark* and enables the receiver to locate the exact area on the map covered by the overlay.
- Sketch the object seen or the information to be transmitted on the tracing paper (the sheet on top of the map) in the exact location it would appear on the map (the sheet underneath the tracing paper).

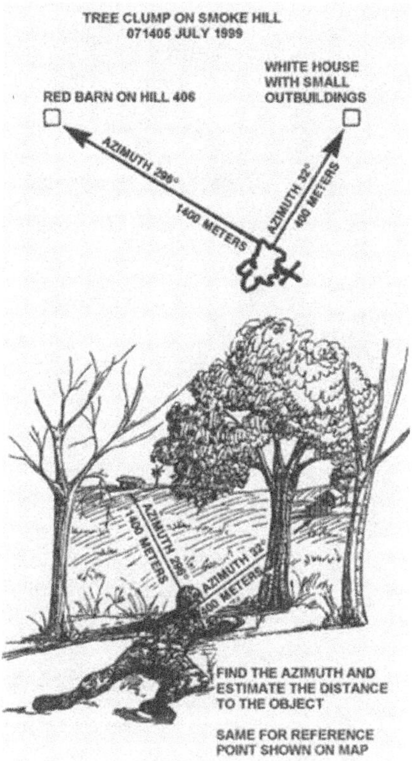

Figure 6-4. Making a Simple Sketch.

Explanatory notes are annotated in the overlay's margin, arrows point to the objectives.

- Indicate with an "X" and an appropriate explanation the position from which the observer saw the object or obtained information.
- Include title and scale of the map from which the overlay was made, date and hour the information was obtained, and signature of the observer on the overlay in the lower right-hand corner.

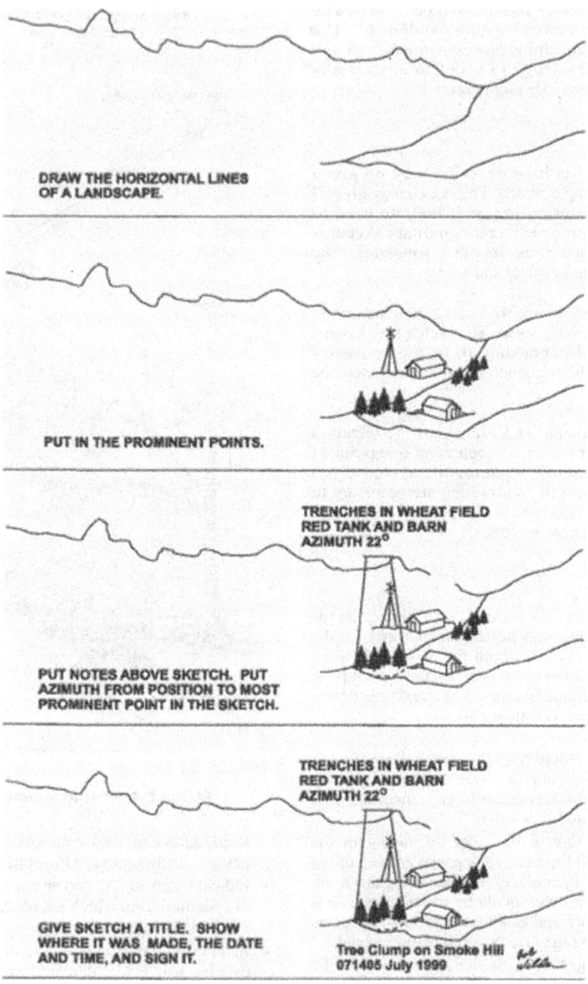

Figure 6-5. Panoramic Sketch.

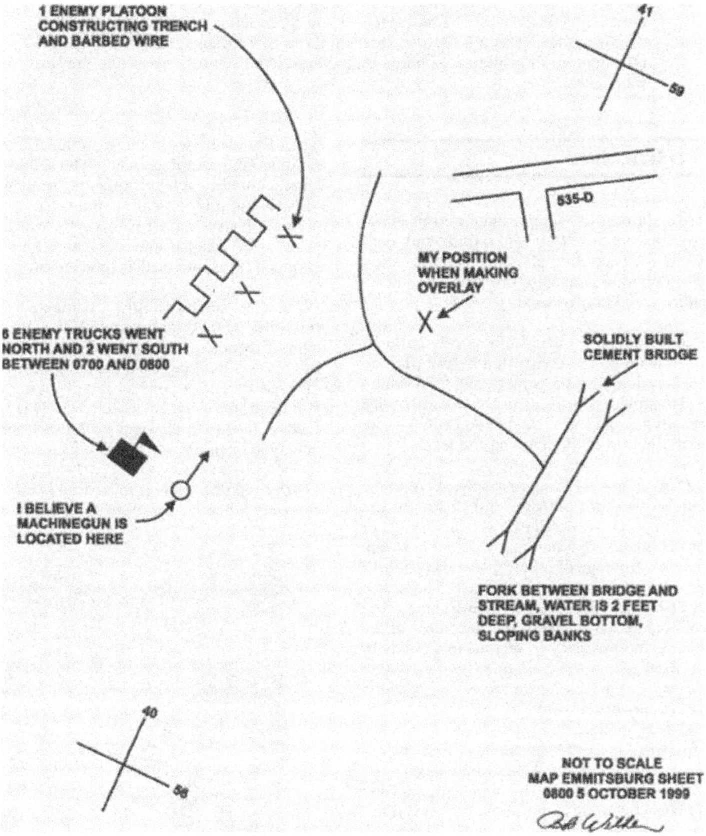

Figure 6-6. Simple Overlay.

CHAPTER 7. SCOUTING FIRE TEAMS

Upon completion of individual training, a Marine adapts this training to the functioning of the unit. The basic tactical unit for scouting is the fire team. Within the fire team, scouts normally work in pairs to furnish security and gather information. Information gathered by the individual scouts is passed to the fire team leader.

7001. POSITIONING

When a rifle platoon in the approach march is not preceded by adjacent units (left, right, front, and rear), it employs its own scouting elements. The scouting element for a platoon is usually one fire team; however, an entire squad may be used.

A fire team used as a scouting element is called a scouting fire team and is controlled by the platoon commander, assisted by the squad leader. A squad leader whose squad is providing the scouting fire teams normally marches near the platoon commander to assist in the control of the scouting fire teams.

A scouting fire team moves aggressively to cover the front of the advancing platoon and to locate the enemy's position(s). It generally moves in a wedge or skirmisher's formation. Normally, a scouting fire team is deployed on a frontage of 50 to 75 meters (10 to 17 meters between each individual scout). The entire squad may be employed to cover a wider frontage. The platoon commander coordinates the movement of the scouting fire team(s) so as to protect the main body of the platoon from enemy fire from points within 400 to 600 meters away, or in close terrain from points within the limits of enemy observation.

Scouting fire team(s) should have enough firepower to overcome resistance from small enemy advanced posts and patrols; the intent is to make enemy riflemen and machine gunners open fire and disclose their positions. Without scouts in advance, the platoon may move into areas where enemy fire may prevent further advance or maneuver and inflict heavy casualties (see fig. 7-1).

Scouting fire teams are covered by the platoon or, when the platoon is masked, the fire team leader ensures individual scouts are maneuvered and coordinated so that the fire team covers its own advance. The fire team leader constantly watches for signals from the platoon commander and remains in visual contact at all times.

The distance between the scouting fire team and the platoon is terrain-dependent. The scouting fire team should not be beyond visibility of the platoon. In open terrain, the platoon commander usually directs the scouting fire team to move by bounds along a succession of locations designated by the platoon commander as intermediate objectives.

Individual scouts should advance as stealthily as possible, while remaining consistent with their mission of reconnaissance to the front, taking advantage of cover without delaying the advance. An occasional glimpse of scouts constantly advancing over a wide front can make the enemy uneasy. It is this activity, and not the target the scouts offer, that may cause the enemy to open fire and disclose its location.

When fired upon, scouts must drop to cover and return fire only when necessary to complete the mission. If

Figure 7-1. Position of Scouts Preceding an Attacking Platoon.

necessary, one scout reports back to the fire team leader who informs the platoon commander.

7002. LOCATING ENEMY POSITIONS

The key terrain of defensive positions are those points that afford extended observation over the ground where the attack must advance. The enemy will place machine guns and infantry to defend critical points.

Members of a scouting fire team preceding an attacking platoon identify the probable enemy infantry and machine gun positions (see fig. 7-2). They use concealment and cover to conduct their advance in order to discover the exact location of enemy positions.

7003. ACTION WITH AN ATTACKING PLATOON

The scouting fire team reconnoiters to the front of the advancing platoon. As soon as the scouting fire team leader indicates the area is secure, the platoon advances and the scouting fire team moves forward. Squads within the platoon advance by bounds; at least one squad is positioned to support the other(s) by fire. Successive positions along the line of advance are selected and designated by the platoon commander as intermediate objectives, and reconnoitered by the scouting fire team before occupation. By conducting proper reconnaissance, surprise by the enemy or movement in the wrong direction may be prevented.

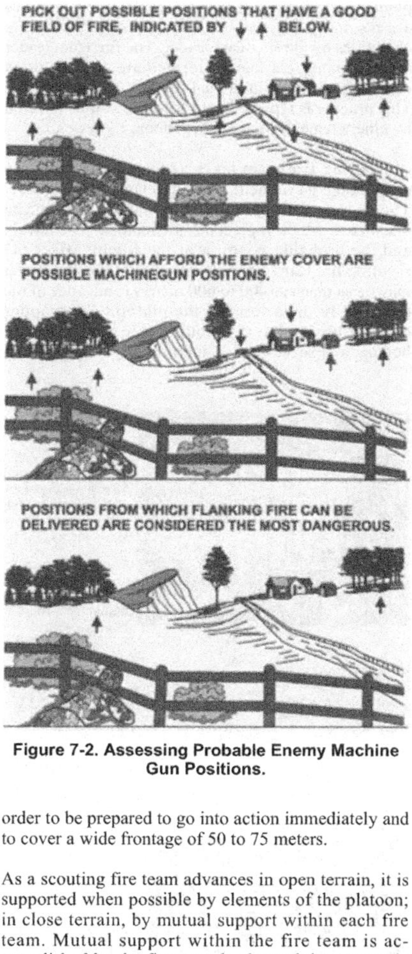

Figure 7-2. Assessing Probable Enemy Machine Gun Positions.

Movement

The distance between the scouting fire team and the front of the main body of the platoon is dependent on the mission, enemy, terrain and weather, troops and support available, and time available (METT-T). In close terrain, such as dense woods, the scouting fire team's movements closely resemble those used for night operations. In approaching houses, woods, and villages, one scout of each pair covers the other while the latter reconnoiters (see fig. 7-3).

A scouting fire team moves forward aggressively to cover the front of the advancing platoon, usually adopting either skirmisher's or wedge formation in order to be prepared to go into action immediately and to cover a wide frontage of 50 to 75 meters.

As a scouting fire team advances in open terrain, it is supported when possible by elements of the platoon; in close terrain, by mutual support within each fire team. Mutual support within the fire team is accomplished by the fire team leader and the automatic rifleman forming a team that supports by fire the advance of the rifleman and assistant automatic rifleman until they reach a location designated by the fire team leader. The rifleman and assistant automatic rifleman then support by fire the movement of the team leader and automatic rifleman. These successive points to which the sub elements of the fire team move are normally designated as fire team intermediate objectives by the fire team leader. The fire team leader sets as many fire team intermediate objectives as necessary to maintain mutual support within the team. This process is repeated until the team can be covered by other elements of the rifle platoon.

A scouting fire team takes advantage of available cover and concealment without delaying its advance. The orders of the platoon commander govern the distance at which it precedes the platoon. The terrain and the probable position of the enemy affect the scouting fire team's distance in front of the platoon. It may be as much as 400 to 600 meters in advance of the platoon. In open terrain, the platoon commander usually directs that the scouting fire team move by bounds to a succession of intermediate objectives. In close terrain or conditions of limited visibility, the scouting fire team is normally ordered to precede the platoon at the limit of visibility, maintaining visual contact with the platoon commander.

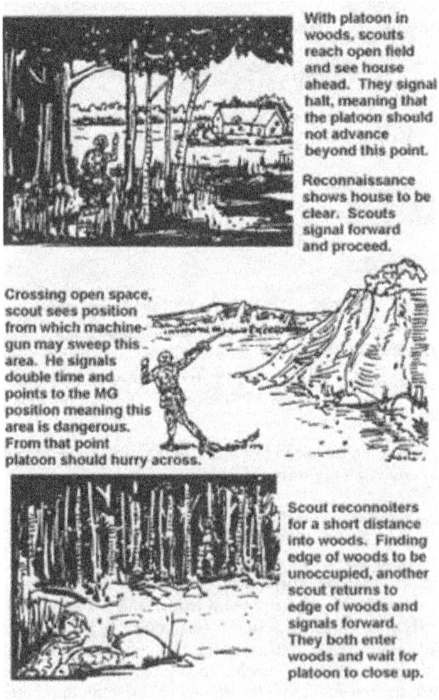

Figure 7-3. Conduct of Scouts Preceding a Platoon.

Action in Woods

When a scouting fire team is directed to advance over open ground to the edge of a woodline, two members of the team, preferably the rifleman and assistant automatic rifleman, reconnoiter inside the woodline while the remainder of the fire team covers them. It is not recommended they separate until finished with their reconnaissance of the far side of the danger area. Both members staying together can cover the same area using a zigzag reconnaissance and they are better equipped to overcome any opposition.

In heavy underbrush and/or poor visibility, the rifleman and assistant automatic rifleman proceed into the woodline together for 50 to 60 meters. The two then separate, searching out either flank to the first high ground or limits of observation, probably 50 to 100 meters (see fig. 7-4 on page 7-4). After the initial search and out posting the limit of advance, the remaining scout signals the fire team forward.

In light underbrush and/or good visibility, the assistant automatic rifleman remains at the edge of the woodline while the rifleman searches the woodline. (see fig. 7-5 on page 7-5). The rifleman searches the woodline in a zigzag pattern, reports back to the assistant automatic rifleman, then moves to an outpost position at the limit of advance. In turn, the fire team leader signals the platoon commander that it is safe for the platoon to move forward.

The fire team leader then moves the remainder of the fire team into the woods, joining up with the forward scout manning the outpost. The scouting fire team occupies and holds a line 50 to 75 meters within the woods and observes toward the direction of movement until the platoon closes up. The scouting fire team leader awaits further word from the platoon commander before moving the team further into the woods.

When directed, the scouting fire team leader moves the team forward until they reach the far edge of the woods. The team is held at the edge of the woods and the fire team leader notifies the platoon commander of the situation. The platoon commander moves the platoon to a position where it can cover the scouting

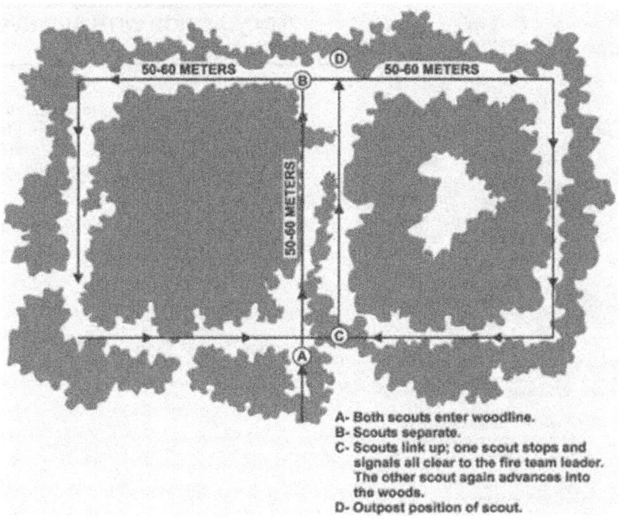

Figure 7-4. Searching Edge of a Woodline (Dense Underbrush/Poor Visibility).

fire team as it exits the woods and directs the team leader to move out and continue the scouting mission.

A scouting fire team passing through woods ahead of its unit maintains a distance allowing visual and oral communications. If an obstacle is encountered, reconnaissance to its front and flanks must be carried out. When advancing along a road or path, scouts precede the platoon to provide necessary protection and to prevent surprise fire on the platoon. When crossing a road or path, they reconnoiter well to the flanks before signaling "all clear" to the platoon.

The scouting fire team will not exit the woods until the arrival of the platoon commander, who will then be given an opportunity to alter the disposition or direction of march. The point where the platoon exits the woods is considered a danger area where the platoon is vulnerable to enemy fire. The scouting fire team is sent ahead to reconnoiter the danger area, as well as the next area to be occupied by the platoon. They signal back whether conditions require a halt, an advance or a quick rush across the open area. The scouting fire team leader must be continually on the lookout for signals from the rear.

Action Under Fire

When a scouting fire team is fired upon, they immediately take cover, locate targets, and return fire. The scouting fire team leader then determines—

- Location of enemy (range and reference points).
- Extent of position (location of flanks).
- Types of positions (obstacles, bunkders, fighting holes, etc.).
- Number of enemy.
- Enemy weapons (machine guns, mortars, tanks, etc.).

The platoon commander assesses the situation as quickly as possible based on the limited information obtained. Usually the platoon commander brings up the remaining squads, sets up a base of fire, and assaults the enemy position. Should the enemy position prove too strong for the platoon, the platoon remains engaged with the enemy as a base of fire until the remainder of the company is committed to clear the enemy resistance.

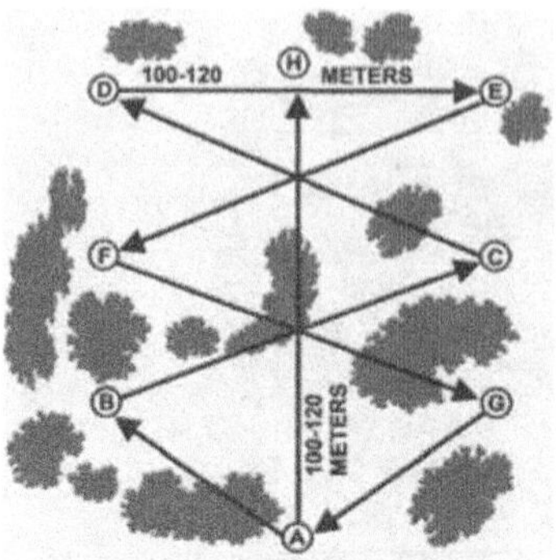

A- Both scouts enter woods. Assistant automatic rifleman remains here and covers forward movement of rifleman.

B through G - Rifleman moves to these positions, stopping and observing before moving to next position.

G to A - Rifleman reports results of his search to the assistant automatic rifleman.

A to H - Rifleman moves to out-post position.

Figure 7-5. Searching Edge of a Woodline (Light Underbrush/Good Visibility).

7004. ACTION WITH AN ENVELOPING UNIT

When a platoon is given the mission to envelop an enemy position, a scouting fire team is employed for protection and reconnaissance in the same way as when the platoon is advancing in the approach march.

PART 2. INFANTRY PATROLLING

CHAPTER 8. FUNDAMENTALS OF INFANTRY PATROLLING

This chapter begins Part 2 and provides basic information about infantry patrols; specifically, their purpose, types, and missions. Infantry patrol training is also addressed and keys to successful patrolling are presented. Subsequent chapters of this part cover patrol organization, preparation, movement, and reconnaissance actions.

8001. DEFINITIONS

A patrol is a detachment of ground, sea or air forces sent out for the purpose of gathering information or carrying out a destructive, harassing, mopping-up or security mission (Joint Publication [JP] 1-02). The mission to conduct a patrol may be given to a fire team, squad, platoon or company.

8002. RELATION OF PATROLLING TO SCOUTING

Each patrol member must be knowledgeable in the principles of scouting and maintain membership of a larger team. To develop the teamwork skills required among the members of a patrol, additional training beyond the basic principles is necessary to become a well-trained scout. A patrol member must respond quickly to the decisions and orders of the patrol leader. There must be complete confidence among all members of the patrol and the confidence that they, as a team, will be successful in their mission.

8003. PURPOSE

A commander must have current information about the enemy and the terrain in order to employ the unit effectively. Patrols are an important means of gaining this information and are used to destroy enemy installations, capture enemy personnel, perform security missions or prevent the enemy from gaining information. Modern warfare places a high premium on effective patrolling because units have larger areas of operations and can be threatened from all directions. As distances between units increase, more patrolling becomes necessary to prevent infiltration by guerrillas or small enemy units, as well as to maintain contact with friendly adjacent units. Active patrolling by numerous small groups is needed to locate the enemy and gather information on the enemy's disposition, strength, morale, and weapons, as well as gather and confirm information about the terrain.

8004. TYPES OF PATROLS

Classification as to Mission

Reconnaissance Patrol

Reconnaissance patrols gather information about the enemy, terrain or resources. Relying on stealth rather than combat strength, they gather this information and fight only when necessary to complete the mission or to defend themselves. The distance covered by reconnaissance patrols varies based on the terrain and mission. The squad is ideally suited for reconnaissance patrol missions because of its relative small size and its experience of working together.

Combat Patrol

A combat patrol is a fighting patrol assigned missions that require engagement with the enemy in combat. Larger and more heavily armed than reconnaissance patrols, combat patrols have a mission to capture enemy documents, provide security, and capture or destroy enemy equipment and installations. Such action is ordinarily followed by a return to friendly positions. Regardless of the mission, the patrol reports any information concerning the enemy and terrain acquired during the accomplishment of the assigned mission. There are four types of combat patrols: raid, contact, ambush, and security (normally conducted by a Marine rifle platoon). A rifle platoon reinforced with crew-served weapons is normally considered the minimum size for contact, economy of force or ambush patrols. In some situations, such as the capture of a small enemy outpost, a rifle platoon could conduct a raid. However, a raid is a complex mission and, due to the organization of a raid force (command, reconnaissance, assault, support, security, and reserve elements), a rifle company is normally the smallest force assigned to a raid.

Classification as to Means of Movement

Foot Patrols

Movement by foot is the most common means; however, there are inherent disadvantages. Foot patrols travel slowly and carry limited quantity and types of equipment and supplies. Range and area coverage is relatively restricted. Foot patrols also have apparent advantages in that they have fewer restrictions as to terrain that can be covered; are more difficult for the enemy to detect; provide thorough coverage within limits of range; and are generally not inhibited by weather.

Motorized Patrols

Where terrain and road networks permit, a motorized patrol overcomes the inherent disadvantages of the foot patrol. Mechanized forces require patrolling units that can keep pace with them. However, motorized patrols are restricted to certain types of terrain, and tend to bypass areas that may be advantageous to and occupied by enemy infantry.

Waterborne Patrols

Waterborne patrols move over seas, lakes, rivers and streams, canals, and other inland waterways. The water is either used as a medium of entry to an objective area or is the actual patrol route. Waterborne patrols are limited by the location of water routes in the terrain and tend to bypass areas that may be advantageous to and occupied by the enemy.

Helicopterborne Patrols

Where terrain is extremely difficult or the enemy situation precludes the use of vehicle or motorized patrols, helicopterborne patrols are a method or means to conduct a patrol.

8005. TRAINING

Training is essential to successful patrolling. Premature and unordered actions by members of the patrol destroy coordination and control. Leaders are trained to issue their orders calmly to inspire confidence and discipline, and to avoid misunderstanding. Patrol members must work together and fight as a team. Training should develop the following skills:

- Expertise in handling individual and special weapons, and familiarity with enemy weapons that may be captured.
- Recognize camouflaged personnel, equipment, and defensive positions; ability to pick up fleeting targets fire the rifle from any firing position.
- Understand fire discipline and, after weapon firing, immediately change position (see fig. 8-1).
- Quick and accurate observation skills, and the ability to recall and transmit clearly and briefly, both orally and in writing.

Figure 8-1. Changing Position After Firing.

- Recognize and quickly respond to improvised signals, visually or by sound.
- Ability to swim with weapon and equipment.
- Use issued or improvised camouflage suits and garnish helmet in order to blend with the surroundings. Smudge face, hands, and any bright surfaces of weapons and equipment with some substance, such as mud or charcoal, to prevent the reflection of light.
- Silence self, equipment, and weapon.
- Use antimalarial and water purification tablets.
- Acclimation to temperature extremes.
- Develop a sense of direction and learn how to follow a course by compass, stars, sun, flow of streams, prominent terrain features, and by observing other natural phenomenon. Learn to determine the distance traveled from a known point and to keep a record of azimuths and the distance traveled on each azimuth (dead reckoning).
- Call for and adjust indirect fire assets.
- Familiarization with all communications assets and the use of field expedient antennae.

8006. KEYS TO SUCCESSFUL PATROLLING

Regardless of the category or means of conducting a patrol, the keys to successful patrolling are—

- Detailed planning. Every portion of the patrol must be planned, all possible contingencies considered.
- Productive, realistic rehearsals. Each phase of the patrol is rehearsed, beginning with actions in the objective area. Similar terrain and environmental conditions are used when conducting rehearsals.
- Thorough reconnaissance. Ideally, the patrol leader will physically conduct a reconnaissance of the route and objective. Photographs and/or maps will be used to supplement the reconnaissance.
- Positive control. The patrol leader must maintain positive control, this includes supervision during patrol preparations.
- All-around security. Security must be maintained at all times, particularly near the end of the patrol where there is a natural tendency to relax.

Chapter 9. Patrol Organization

Organizing a patrol is a two-step process: the general organization of the entire patrol and the task organization of various patrol elements. Normally, the nature of patrolling does not permit long preparation periods and rehearsals to specifically build a unit for each mission. Accordingly, the patrol leader must combine unit integrity considerations with proven concepts of patrol organization.

9001. GENERAL ORGANIZATION

The patrol leader establishes a patrol headquarters and elements to accomplish the mission.

The headquarters is composed of the patrol leader and the personnel who provide support for the entire patrol, such as a forward observer, corpsman, and radio operator.

The major subdivisions of reconnaissance and combat patrols are elements. The existing infantry structure (squads and fire teams) is reinforced as required. For example, a reinforced platoon tasked to conduct a combat patrol that will raid an enemy outpost could be organized as follows:

- Platoon headquarters (command element):
  - Patrol leader (platoon leader).
  - Assistant patrol leader (platoon sergeant).
  - Navigator.
  - Radio operator (company tactical net).
  - Radio operator (patrol tactical net).
  - Corpsman.
- The first squad (security element) provides security en route to the objective area (the point, flank security, and rear security) and at the objective area (the flanks and the objective rally point).
- The second squad (support element) supports by fire for the attack, covering fire for the withdrawal, and supporting fires to cover the crossing of danger areas.
- The third squad (assault element) provides the assault force to attack and seize the objective; searchers to clear the objective; pacers, compass man, navigator, and the assistant patrol leader en route and back from the objective area.

Any attachments a patrol may have (i.e., demolition team, scout snipers, and machine gun squad) will be added to the element that supports its function. For example, the demolition personnel should go with the unit conducting the attack, and scout snipers and machine gunners should stay with the support squad.

9002. TASK ORGANIZATION

The preceding paragraphs described the elements necessary for a patrol to accomplish its mission. These elements reflect the internal functions or tasks required for the patrol to succeed. Depending on the METT-T, there are various methods of grouping these elements together. Task-organization is the further subdivision of patrol elements into teams that are required to perform essential tasks. In creating teams, unit integrity of infantry units should be maintained.

The patrol is organized so each individual, team, and element is assigned a specific task, but capable and prepared to perform other tasks. This may not be possible for certain specialist tasks requiring a trained technician.

Chapter 10. Patrol Preparation

For a patrol to succeed, all members must be well trained, briefed, and rehearsed. The patrol leader must have a complete understanding of the mission and a thorough understanding of the enemy and friendly situations. The patrol leader should make a complete reconnaissance of the terrain to be covered (either visual or map), and must issue an order to the patrol, supervise preparations, and conduct rehearsals.

10001. MISSION

The mission assigned to a patrol must be clear and oriented toward one objective with a specific task and purpose. More than one primary objective or indefinite missions invites confusion, casualties, and failure.

10002. FACTORS INFLUENCING PATROL SIZE

The circumstances under which patrolling by infantry units is conducted make it necessary for combat patrols to be able to fight offensively, security patrols to defend themselves, and reconnaissance patrols to move quickly and only fight if necessary to break contact or defend themselves. The size of a patrol depends on METT-T.

Generally, a patrol should contain the least number of members needed to accomplish the mission. Combat missions ordinarily require larger patrols than reconnaissance missions.

10003. COMMANDER'S DUTIES

Determining Patrol Requirements

The need for conducting patrols derives from the commander's stated mission (issued by higher level commander) and other specified and implied tasks (secondary or supporting efforts necessary to accomplish the stated mission). This requires analyzing the unit's mission and determining the necessary reconnaissance and/or combat tasks that must be performed. By considering the mission, along with time available, the commander develops the overall concept of operations to include the patrol plan and the specific tasks for each patrol.

Assigning Units

When assigning patrol missions, maintaining the integrity of the existing unit while considering the skills and experience of the unit and its leader are critical factors to the infantry commander. To provide operational depth and equitable apportionment of hazardous assignments, the commander ensures that each of the subordinate leaders and units develops the skills and experience necessary to conduct successful patrols. Assignment of patrol units must consider the commander's concept of operations as a whole and the plans for subsequent employment of assigned forces after completion of the assigned patrolling mission.

Providing Adequate Time

The commander must allow the patrol sufficient preparation time by completing the mission analysis, estimate, and preparation of orders. The commander should use warning orders to alert subordinates to possible requirements and afford them the opportunity for concurrent planning.

Providing the Patrol Leader Information

The following information should be provided:

- A simple, straightforward explanation of the mission, particularly for night patrols.
- General routes (defined by checkpoints) or exact routes (defined by avenues of approach or other terrain features) to follow.
- Enemy composition, disposition, and strength.
- Location and activities of friendly troops.
- Outposts or other security elements through which the patrol is to pass.
- Terrain conditions.
- Missions and routes of other patrols.

- Time patrol is to depart and return.
- Method of reporting information while on patrol (radio, messenger), place where messengers are to be sent, and place where the patrol leader is to report upon completion of the patrol.
- The challenge and password to be used for exiting and reentering friendly lines.
- Special instructions such as locations to be avoided and essential elements of information higher headquarters is seeking.
- Report barrier/obstacle plan of friendly units if known; support available from friendly unit (e.g., medical evacuation [MEDEVAC], react force).
- Fire support available.

Providing Special Skills and Equipment

There will be situations when the unit assigned to conduct a patrol does not have the necessary technical skills or equipment organic to it to successfully accomplish the patrol mission. In these cases, additions (attachments) are made to the unit, such as—

- Machine gun and/or shoulder-launched, multipurpose assault weapon (SMAW) team and/or squad.
- Forward observers (mortar/artillery).
- Radio operator(s).
- Combat engineers.
- Tracked vehicle crewman to assess the ability to traverse the terrain.
- Corpsman.
- Other personnel (snipers, translators) as required.

Providing Miscellaneous Support

The commander must ensure that the patrol leader is provided with the food, water, ammunition, radios and batteries, maps, special clothing, and any other items required by the unit (includin attachments) needed for the mission. Post-patrol support such as debriefings by intelligence personnel must also be planned.

Reviewing the Patrol Leader's Plan and Preparations

Once the patrol leader receives the mission, conducts visual and/or map reconnaissance, and develops the plan, the commander confirms the patrol leader's understanding of the mission and plan for accomplishing it. This discussion between the patrol leader and commander ensures that the patrol leader understands the commander's intent and is properly prepared to provide the patrol leader an opportunity to ask for clarification or additional support, if required.

Debriefing the Patrol

Upon return of the patrol, the commander receives the patrol report at a debriefing attended by the patrol leader and all patrol members. The debriefing should be conducted as soon as possible following the patrol's return, while information is still fresh in the minds of the patrol members. A patrol report, based on the information collected during the debrief, is generated and forwarded to the next higher com-mander.

10004. PATROL LEADER DUTIES

The patrol leader organizes and prepares the patrol by using the six troop-leading steps to make the best use of resources available. These steps are to—

1. Begin planning.
2. Arrange for reconnaissance and coordination.
3. Make reconnaissance.
4. Complete the plan.
5. Issue the order.
6. Supervise.

Begin Planning

The patrol leader begins by evaluating all factors affecting the mission. He looks for possible courses of action that lead to a decision, and then transforms this decision into an order. The first step includes making an initial assessment and decision on using available time, issuing a warning order and initial preparatory tasks, and initiating his estimate.

The initial planning effort assesses the time, assistance, and information available, and plans the proper use of each. Time allowances include reconnaissance; completion of the estimate and order; troop preparation; and such briefings, rehearsals, and inspections as required before beginning the patrol.

The patrol leader reviews the mission and the attachments and/or support available and decides what preparatory efforts must begin immediately. Proper use of subordinates to manage these initial tasks during this period reduces preparation time and frees the patrol leader for proper planning and reconnaissance. A warning order is issued using a modified five-paragraph order format (situation, mission, execution, administration and logistics, and command and signal [SMEAC]) as a checklist. (Refer to app. A for specific information that a patrol warning order should contain within each checklist item.

Once the initial preparations are set in motion, the patrol leader begins estimating by analyzing the mission; considering the friendly, enemy, and operating environments; considering each course of action available against what the enemy might do; comparing the courses of action in terms of mission accomplishment, capabilities, and probable casualties; and choosing one that becomes the basis for concept and order. While the patrol leader should use and organize notes, estimates must be done quickly and accurately, particularly for immediate situations. To organize thinking, the patrol leader uses METT-T, which consists of:

- Mission—the mission assigned to the patrol and how it relates to the mission of the commander who is sending the patrol.
- Enemy—what is known or suspected of enemy presence and capabilities, habits and characteristics, and fighting techniques.
- Terrain and weather—including ground, vegetation, drainage, weather, and visibility.
- Troops and support available—friendly situation and support available.
- Time available—the constraints and impact of time on preparation and mission accomplishment.

The estimate begins with mission analysis, which is the most important part of the entire planning process. Here, the patrol leader considers the specified tasks of the mission assigned and identifies other significant actions (specified and implied tasks) that must be undertaken to accomplish the stated mission. These tasks are arranged in sequence of accomplishment. The tasks and sequence create the framework for developing courses of action for the patrol concept of operations.

Arrange for Reconnaissance and Coordination

The patrol leader arranges a personal reconnaissance to observe as far forward as possible and also coordinates with the appropriate commanders for the patrol's "passage of lines" (see paragraph 11001 for complete definition) and supporting fires. The patrol leader also coordinates with other patrol leaders who may be operating in the same or adjacent areas and requests that the commander assigning him the patrol mission coordinate the patrol action with adjacent commanders, local security, and night defensive fires, as appropriate. The patrol leader may delegate any or all of these arrangements to the assistant patrol leader if the patrol leader requires the time for planning.

Make Reconnaissance and Complete the Estimate

The patrol leader uses personal reconnaissance to answer questions that arise from the map reconnaissance and METT-T evaluation. Specific points include passage points, lanes through obstacles, locations of friendly listening posts and observation posts, possible approach and return routes, enemy positions (if any), and intermediate observation points on the way to the objective.

In selecting approach and return routes, the patrol leader chooses routes that best use concealment and avoid opposition and obstacles. To lessen the chances of ambush by the enemy, the return trip is planned along a different route. In addition to personal reconnaissance and review of the map and aerial photographs, the advice of other patrol leaders who already are familiar with the terrain and the objective area should be considered.

After compiling information about the situation and possible time constraint, the patrol leader completes an estimate. The first step is developing courses of action, each of which will provide for movement to the objective area, mission accomplishment, and the return, based on the tasks and their sequencing identified in mission analysis. While the eventual concept of operations is presented in order of occurrence, the patrol leader must develop the courses of action by either backward or forward planning. In situations where the objective is well defined and there is sufficient information to plan the action for mission accomplishment (reconnaissance or combat), the patrol leader begins the scheme for accomplishing the mission at the objective and then, planning backwards, considers the options for getting there and back.

The following sections on movement to and return from the objective area, reconnaissance missions, and combat missions discuss methods and options available to the patrol leader in developing the courses of action. The principal variables between courses of action will be who, where, and how in the following:

- Patrol task organization.
- Routes to the objective area.
- Observation point(s) (reconnaissance patrol), ambush site, form of maneuver, type of ambush (combat patrol), fire support plan.
- Return routes.

The patrol leader then mentally considers the progress of each course of action (a map or simple sketch is a useful aid) against expected and unexpected enemy action. By comparing the options against each other and prospective enemy opposition, the patrol leader chooses the course of action that has the best chance of success. Includes in this mental preview process is the time to determine the patrol plan for unexpected contingencies (enemy attack/counterattack, casualty handling). These contingency actions, together with the selected course of action, become the patrol concept of operations.

Once the patrol leader determines the scheme of maneuver, the fire support required to accomplish the mission is addressed including the fire power organic to the patrol and what additional indirect fire support will have to be provided by other units. When planning for indirect fire support, the patrol leader considers the following questions:

- Will artillery, mortar, or close air support be required at the objective area (combat patrols)?
- What artillery and mortar targets exist along the routes to and from the objective area that can be employed by the patrol if it encounters the enemy during movement (reconnaissance patrols and combat patrols)?
- What additional fire support will be required to cover the patrol's movement from the objective area back to the friendly area once the enemy is aware of the patrol's actions at the objective area (combat patrols)?

The effect that casualties have upon the patrol depends upon many factors. Generally, more casualties can be expected in a combat patrol than in a reconnaissance patrol. A patrol may continue on to the objective carrying its casualties, send them back with a detail of Marines, abort the mission and return the entire patrol with the casualties, or call their parent unit for assistance.

Some factors that determine what action the patrol leader takes are: patrol's mission; unit's standing operating procedure for handling wounded; number of casualties and nature of their injuries; availability of aid, helicopters or other means of casualty evacuation. Helicopter evacuation should only be used for the most serious casualties. For infantry units conducting patrols in proximity to the enemy, helicopter evacuation of casualties may compromise the patrol's mission and force the patrol to return to friendly positions before the mission is completed.

The patrol leader determines the requirement for nuclear, biological, and chemical (NBC) defense equipment. Gas masks should always be carried due to the availability of riot control agents (RCAs) to the enemy. If chemical or biological agents have been employed in the area that the patrol must pass through, protective garments will have to be worn by patrol members for part of or the entire patrol. Wearing extra clothing and carrying extra equipment affects the speed of the patrol's movement. A contingency plan for post patrol decontamination must be developed.

Complete the Plan

At this point, the patrol leader has completed the basic thinking necessary for accomplishing the assigned mission. The patrol leader prepares the patrol order to spell out the details, assign tasks to subordinates, and explain the entire endeavor for ease of understanding by the other members of the patrol.

Prepare the Order

The patrol leader's order contains more detailed information than discussed in the warning order. Orders follow the prescribed five-paragraph order format but contain greater detail. The patrol order is a modified 5-paragraph order; the major modifications are to paragraphs 3a and 3c. The format for the order is contained in appendix B.

Issue the Order

The patrol leader asks for a status report on the initial preparatory tasks assigned to subordinate leaders and specialists when the warning order was issued. When the patrol leader has completed planning and initial preparations have progressed to the point where the patrol order may be issued, the members of the patrol are assembled. Roll call is taken to ensure all patrol members are present, then the prepared order is issued.

This will be the only opportunity for the patrol leader to issue detailed instructions. The mission, in particular, must be unmistakably clear so that once the patrol is committed, all subordinate leaders can act with unity of purpose.

Whenever possible, the patrol leader should have a Marine, such as the navigator, build a terrain model using dirt, sand, twigs, etc., explaining the concept of operations for movement to the objective area, actions at the objective area, and the return. Terrain models provide patrol members with a clear and simple layout of the area of operations and key terrain.

Supervise

Inspections and rehearsals, vital to proper preparation, are conducted even though the patrol leader and patrol members are well experienced in patrolling. Inspections determine the patrol's state of physical and mental readiness.

The patrol leader inspects before rehearsals to ensure completeness and correctness of uniform and equipment. The following areas are checked:

- Camouflage.
- Identification tags, Geneva Convention cards.
- Prescribed equipment, weapons, and ammunition are available and serviceable.
- Tape and other items are used to "silence" equipment (prevent noise produced during movement).
- Items that could provide information to the enemy (e.g., letters and papers) remain behind.
- Unnecessary equipment and excess weight remain behind.

The patrol leader questions each patrol member to ensure the following is known:

- The mission, planned routes (primary and alternate), and the fire support plan of the patrol.
- The individual's role: what to do and when to do it.
- What others are to do and how their actions impact.
- Challenges and passwords, codes, reporting times, radio call signs, frequencies, and any other pertinent details.

There is usually a period of time between final rehearsal and departure. The patrol leader reinspects just before departure to ensure all equipment is still in working order and the unit is ready to embark on the mission.

Rehearsals ensure the operational proficiency of the patrol. Plans are checked and needed changes are made. The patrol leader verifies the suitability of equipment. It is through rehearsals that patrol members become thoroughly familiar with the actions to take during the patrol.

If the patrol will operate at night, both day and night rehearsals are conducted. Terrain similar to that over which the patrol will operate is used. All actions are rehearsed. If time is limited, the most critical phases are rehearsed. Action at the objective is the most critical phase of the patrol and is always rehearsed.

An effective method is to talk the patrol through each phase, describing the actions and having each member perform individual duties. When satisfied, the patrol leader walks the patrol through all phases of the patrol using only the signals and commands to be used during the actual conduct of the patrol. Rehearsals continue until the patrol is thoroughly familiar with the plan. The rehearsal is also used to test the soundness of the patrol order and patrol organization.

After the rehearsal, the patrol leader makes final adjustments to the plan and patrol organization based on what was learned during the rehearsal and from other sources, such as the S-2 and adjacent patrols. When this is completed, the patrol leader issues final instructions to subordinate leaders noting any changes made in the patrol organization or plan. While the subordinate leaders are briefing the remainder of the patrol members, the patrol leader reports to the commander stating that the patrol is ready to begin the mission. The patrol leader also coordinates the location and time that the patrol can test fire all weapons prior to departure.

Chapter 11. Movement to and Return from the Objective Area

This chapter provides guidance to patrol leaders for movement to and return from the objective area. Action in the objective area depends on whether the patrol is assigned a reconnaissance mission, combat mission or security mission. Chapters 12 and 13 provide detailed guidance on these types of missions.

11001. Passage of Lines

A passage of lines is an operation in which a force moves forward or rearward through another force's combat positions with the intention of moving into or out of contact with the enemy. (JP 1-02)

During the initial preparation for the patrol, the patrol leader selects a patrol assembly area and reconnoiters the area of passage designated by the commander. In coordination with the unit commander responsible for the area of passage, the patrol leader identifies gaps or lanes in minefields and wire obstacles and locates local security elements through which the patrol will pass. The patrol leader also checks the route from the patrol assembly area to the passage point or contact point where the patrol will depart friendly lines. If possible, both the route to the passage point and the route through the frontlines should be concealed from the enemy's view.

The patrol leader also reconnoiters the area for return passage of lines and coordinates with the unit commander responsible for the area of passage for passage points and lanes as necessary. The patrol leader observes these points from the direction that the patrol will use upon return to friendly lines, if possible, to aid in recognition upon return. The patrol leader provides the forward unit with information about the size of the patrol, general route, and expected time of return. The manner of challenge and recognition of the returning patrol should be coordinated in detail.

Upon return to friendly forward local security squads and/or frontlines, the patrol leader leaves the patrol in a covered position and moves forward with a radio operator and at least one Marine for security to make contact with the friendly unit in the manner agreed. After contact is made and recognized, the patrol leader rejoins the patrol and takes them to the passage point, personally checking in each member.

11002. Organization for Movement

The patrol's task organization establishes the elements and teams needed to accomplish the mission in the objective area and to and from the objective. The patrol leader determines the formation(s) in which the patrol moves to the objective area.

Formations

The proper use of patrol formations is critical to the patrol's success. The squad and fire team formations described in FMFM 6-5 (proposed MCWP 3-11.2), *Marine Rifle Squad*, also apply to infantry patrolling formations. Because the movement of the patrol must be concealed from the enemy, the patrol normally moves through terrain that provides concealment. Control of the patrol in this type of terrain is difficult; thus the column formation, which is easily controlled, is normally used. However, as various types of terrain are encountered, the patrol leader uses the same considerations in determining the appropriate formation used in other infantry operations.

The standard squad and fire team formations are adaptable to a patrol. The patrol may change formations en route to match the situation and terrain. The patrol leader may have to sacrifice some control for better dispersion or give up some speed for greater stealth and more security. Other considerations include—

- Visibility, weather, terrain, and vegetation will influence dispersion and control of individuals and units. These factors may also affect the enemy; if visibility is good for the patrol, it is also good for the enemy. Two pieces of luminous tape worn on the back of the collar will aid in control and movement on dark nights. The collar is turned down when near the enemy. The tape can also be worn on the back of the cap, but should be covered or removed when near the enemy.

- Preserving the integrity of fire units (fire teams and, if attached, machine-gun teams) is of primary importance. If team members are detached from a machine-gun team, the unit integrity is lost and effectiveness is reduced. The patrol leader must also position fire units so as not to mask their fires.
- All-around defense of the patrol must not be sacrificed under any circumstances. The conventional squad and platoon formations provide adequate firepower in any direction required. When attachments are made to the patrol, the attachments are positioned within the formation to enhance the firepower of the patrol. If a fire unit, such as a machine-gun team or squad, is attached, it is incorporated into the all-around defense of the patrol by modifying the conventional formation (see fig. 11-1). It is permissible to employ machine-gun teams individually during movement. However, assault rocket squads and teams should be employed primarily as rifle fire units for movement. Firing the SMAW or AT-4 from within a patrol formation can present a serious back-blast danger to patrol members and should only be fired on the patrol leader's direction.
- Time allotted for mission accomplishment is also a major consideration. In selecting the formations, the patrol leader must consider the speed of the movement required to meet the time constraints (if any) imposed on the patrol. If required to meet a time schedule, a formation that permits rapid movement should be used. Speed, however, must never be permitted to force the patrol leader to make rash tactical decisions.

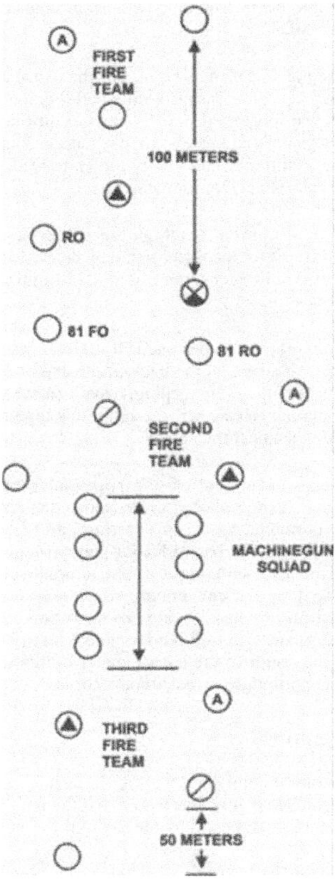

Figure 11-1. Squad Patrol with Attached Mortar Forward Observer Team and Machine Gun Squad.

Exercise of Control

The patrol leader is positioned for best control over the patrol. The assistant patrol leader moves at or near the rear of the patrol. Other subordinate leaders move with their elements. All patrol members assist by staying alert and passing on signals and orders. A signal to halt may be given by any patrol member, but the signal to resume movement is given only by the patrol leader.

Arm-and-hand signals are the primary means of communication within a patrol and should be used exclusively when near the enemy. All members must know the standard infantry signals (refer to FMFM 6-5), as well as any special signals required, and be alert to receive and pass them to other members.

The patrol leader should speak just loudly enough to be heard. At night, or when close to the enemy, the patrol leader halts the patrol and has subordinate leaders come forward. They speak in a low voice and then pass the information to their subordinates by moving from member to member.

Radios provide a means of positive control within a large patrol; however, radios should be used only when arm-and-hand signals or face-to-face contact between the patrol leader and subordinate leaders is impractical. When close to the enemy, words are spoken into the microphone with hands cupped over it in a low voice.

Other sound signals may be used if the patrol leader is sure they serve the purpose intended. Planned sound signals are rehearsed before starting on the patrol. Sound signals used must be simple, natural sounds that are few in number and easily understood. Bird and animal calls are seldom satisfactory.

Infrared equipment, such as the sniper scope and infrared filters for the flashlight, may be used as means of sending and receiving signals and maintaining control at night.

Luminous tape may be used to assist in control at night. Small strips on the back of the cap or collar of patrol members aid in keeping visual contact with the front member. However, the luminous tape must be covered when near the enemy.

An important aspect of control is accounting of personnel, especially after crossing danger areas, enemy contacts, halts, and exiting and re-entering friendly lines. The patrol leader may arrange for the last member to send up the count automatically after crossing danger areas, enemy contact, and halts. In large patrols or when moving in a formation other than a column, subordinate leaders check members and report the count to the patrol leader by the quickest method appropriate to the circumstances.

Navigation

One or more members are assigned as navigators for the patrol. Their function is to assist the patrol leader in maintaining direction by the use of the compass.

The patrol leader assigns at least two members as pacers to keep track of the distance from point to point. The average of their count is used to approximate the distance traveled. Pacers are separated so they do not influence each other's count. Pacers must know how to convert their own paces to meters.

The route is divided into increments, with each increment starting at a recognizable point on the ground. The pacers begin their counts from zero at the beginning of each leg. This makes the pace count easier to keep and checks for accuracy.

The pace count is sent forward when the patrol leader turns to the member behind and in a low voice says, "send up the pace" or uses the arm-and-hand signal of tapping his boot. This signal is passed to both pacers, who in turn send up the pace count in meters; for example, "two-hundred" or "one-seven-five."

Patrol members must understand that the counts of both pacers are sent forward. The patrol leader must know the counts of both men in order to check them.

Security

The patrol is dispersed consistent with control, visibility, cover, and concealment. Scouts are employed to the front, flanks, and rear of the patrol to provide security. For the patrol members traveling in the main body of the patrol, areas of responsibility are assigned to the front, flanks, and rear. Scouts are the eyes and ears of the patrol leader. They move when and where directed by the patrol leader and maintain contact with the patrol leader at all times, except when momentarily obscured by vegetation or other terrain features.

Front

Small patrols (squad size) may employ from one scout up to a fire team as the point, depending on the enemy situation, terrain, and route being followed by the patrol. Normally, squad-size patrols will use two scouts as the point. The point is responsible for investigating the route of advance immediately to the front of the patrol.

The point moves as far ahead of the patrol as visibility and terrain permit. When visibility is good, the point may precede the main body by as much as 100 meters. The point travels right and left ahead of the patrol, searching the area over which the patrol will pass.

The point maintains direction by knowledge of the general route to be followed and visual contact with the patrol leader. The patrol leader or the navigator ensures that the point is proceeding correctly.

The point, which stays far enough ahead of the patrol to provide security, is not a trail breaker for the patrol. If the point loses contact with the patrol, the point waits for the main body to catch up or moves rearward if contact is not quickly regained.

One of the navigators may be positioned with the point. One or more members works as the point while the other is the navigator.

Flank

Flank security for a patrol of squad size or less may be provided by using one or two members on either flank. If two scouts are assigned to a flank, one is positioned to observe the patrol leader and the other works farther out from the patrol. The scout who must observe the patrol leader remains within a maximum distance of 100 meters. The scout farther out remains in sight of the inside scout but normally does not move more than 20 or 25 meters away and remains prepared to relieve flank security regularly. Moving through dense woods or jungle may render the use of flank security impractical because of reduced visibility. In such cases, it moves with the patrol itself, but maintains observation to its assigned flanks.

Rear

A small patrol normally has only one rifleman assigned as rear security. An interval between the member assigned as rear security and the last Marine of the patrol is maintained at the limit of visibility, up to 50 meters. This member maintains rear security for the patrol by constantly observing to the rear.

Halting

Speed of movement is slower at night than in day patrols and reduces the danger of a Marine becoming separated from the patrol. The patrol occasionally halts to observe and listen for enemy activity; this is called a security halt. Upon signal, when reaching a danger area and periodically throughout movement en route, every member freezes in place, remains quiet, observes, and listens. It may be necessary to call a security halt just after departing friendly areas and just before entering friendly areas.

The patrol may halt briefly to send a message, eat, rest, check direction, or make a reconnaissance. The area selected should provide adequate concealment and cover, as well as favor the defense. All-round security is established and the patrol leader ensures all members move out when the patrol resumes movement. (For extended halts, see paragraph 11005.)

Infiltration

The disposition of enemy forces may sometimes prevent a patrol from entering the enemy occupied area as a unit; however, pairs of scouts or fire teams may slip through without being discovered. (Refer to FMFM 6-5 for infiltration techniques and procedures).

11003. CONTROL MEASURES FOR MOVEMENT

Checkpoints

A checkpoint is a predetermined point on the surface of the Earth used as a means of controlling movement, a registration target for fire adjustment or reference for location (JP 1-02). Checkpoints are a means of control between the parent unit and the patrol. These locations are decided upon and coordinated before the patrol leaves, so that both the patrol members and parent unit will know the patrol's location when it reports in. The parent unit can follow the progress of the patrol without transmitting coordinates over the radio that the enemy could monitor.

Rally Points

A rally point is an easily identifiable point on the ground where units can reassemble and reorganize if they become dispersed (MCRP 5-12C). It should provide cover and concealment, be defensible for at least a short time, and be easily recognized and known to all patrol members. All rally points are considered and identified as tentative rally points until they are occupied, found to be suitable, and designated as rally points by the patrol leader.

Initial Rally Point

The initial rally point is within friendly lines where the patrol can rally if it becomes separated before departing the friendly area or before reaching the first en route rally point. It may be the assembly area where the patrol waits while the patrol leader contacts the last friendly position through which the patrol will pass. The location of the initial rally point must be coordinated with the forward unit commander.

En route Rally Points

En route rally points are between the initial rally point and the objective rally point; and from the objective rally point back to the point where the patrol reenters friendly lines. They are determined as the patrol passes through a likely area that is suitable for a rally point.

Objective Rally Point

The objective rally point is located nearest the objective where the patrol makes final preparations prior to approaching the objective. It also serves as a location where the patrol reassembles after completing actions on the objective. The objective rally point must be suitable to accommodate those activities accomplished prior to actions on the objective. This position must provide the patrol concealment from enemy observation and, if possible, cover from enemy fires. It may be located short of, to a flank or beyond the objective. It should be out of sight, sound, and small-arms range of the objective area. The patrol leader's reconnaissance of the objective is made from this position; it is the release point from which patrol elements and teams move to the objective to accomplish the mission.

Rally Point Selection

The patrol leader selects likely locations for tentative rally points during reconnaissance or map study. A tentative rally point must be confirmed and announced after examination proves suitability. A tentative initial rally point and a tentative objective rally point are always selected and identified in the patrol order. If necessary, the patrol leader selects additional rally points en route as suitable locations are reached. When the patrol reaches a danger area that cannot be bypassed, such as an open meadow or stream, the patrol leader selects a rally point on both the near and far side. If good locations are not available, the patrol leader designates the rally points in relation to the danger area. The patrol leader will say, for example, "50 meters this side of the trail" or "50 meters beyond the stream."

Rally Point Use

The initial rally point and en route rally points are designated to enable the patrol to reassemble if it is unavoidably separated or dispersed. Identifying features are pointed out. The patrol leader ensures that the information is passed to all patrol members. When crossing a danger area, a near side rally point and a far side rendezvous point are designated. These rally points should only be used when all other methods of retaining control of the patrol have failed. The success of the patrol may be jeopardized if it is dispersed and cannot rally expeditiously.

If the patrol has left the friendly area and becomes dispersed, patrol members return to the last designated rally point (the initial or an en route rally point) unless the patrol leader gives other instructions.

As previously noted, the patrol leader selects two rally points at the near and far sides of danger areas that cannot be bypassed. If the patrol becomes separated or dispersed at a danger area, and there has been no enemy contact, the patrol should reassemble at the rally point on the far side of it. If, however, the patrol is separated or dispersed at a danger area as a result of enemy contact, members who have already crossed the danger area assemble and reorganize at the rally point on the far side; members who have not crossed assemble and reorganize at the rally point on the near side. In the absence of the patrol leader and unless directed otherwise, the senior member at the rally point on the near side takes charge, attempts to move the rallied patrol members to the rally point on the far side, and rejoins the remainder of the patrol.

Rally Point Actions

The patrol leader plans the actions to be taken at rally points and instructs the patrol accordingly in the patrol order. Planned actions at the initial rally point and en route rally points must provide for the continuation of the patrol as long as there is a reasonable chance to accomplish the mission. Plans for actions at rally points should provide for—

- Recognition signals for assembly at rally points.
- Minimum number of members and maximum amount of waiting time required before the senior member at the rally point moves the rallied patrol members onward toward the objective or returns to friendly lines.
- Instructions for patrol members who find themselves alone at a rally point.

11004. PRECAUTIONS AT DANGER AREAS

A danger area is any place where the patrol is vulnerable to enemy observation or fire (open areas, roads, trails, and obstacles such as barbed wire, minefields, rivers and streams, and lakes). Any known or suspected enemy position the patrol must pass is also a danger area. The patrol leader plans for crossing each danger area and includes these plans in the order.

The patrol reconnoiters the near side of a danger area first, then the patrol leader sends scouts to reconnoiter the far side. Once the scouts report that the far side is clear of the enemy, the remainder of the patrol crosses the danger area. As each individual or group crosses the danger area, they are covered by those remaining and by those who have successfully crossed. Enemy obstacles are avoided since they are usually covered by fire.

In crossing a river, the near bank is reconnoitered first; then the patrol is positioned to cover the far bank. Scouts are sent across to the far bank. After the far bank has been reconnoitered and the scouts report that it is clear of the enemy, the patrol crosses as rapidly as possible. This may be done individually or in pairs. If crossing the river requires swimming, the patrol uses improvised rafts to float equipment, weapons, and ammunition across. (Refer to MCRP 3-02C, *Water Survival Handbook.)*

A road or trail is crossed at or near a bend or where the road is narrow. Observation is restricted and, if the enemy is present, the patrol is exposed as short a time as possible. The near side is reconnoitered first, then scouts are sent across to reconnoiter the far side. This includes reconnoitering the tentative rally point on the far side. Once the scouts report "all clear," the remainder of the patrol crosses rapidly and quietly.

If the patrol must pass close to an enemy position, it takes advantage of battlefield noises to cover the sounds of movement. If supporting fires are available, the patrol leader can call for them to divert the enemy's attention as the patrol passes.

11005. HIDE

When a patrol is required to halt for an extended period in an area not protected by friendly troops, the patrol moves into a location which, by the nature of the surrounding terrain, provides passive security from enemy detection. Such an assembly area is termed a *hide*. To establishment a hide—

- Cease all movement during daylight hours to avoid detection.
- Hide the patrol for an extended period while the patrol leader conducts a detailed reconnaissance of the objective area.
- Rest and reorganize after extended movement.
- Reorganize after a patrol has infiltrated the enemy area in small groups.

The patrol leader's plan must include tentative hide locations when the patrol's mission dictates an extended halt within enemy areas. These tentative locations must be confirmed by actual ground reconnaissance prior to occupation by the patrol. The plan for a hide includes both passive and active security measures.

Passive security measures are—

- Avoid built-up areas.
- Select an area remote from all human habitation.
- Avoid known or suspected enemy positions.
- Avoid ridge lines, topographic crests, valleys, lakes, and streams.
- Avoid roads and trails.
- Avoid open woods and clearings.
- Select areas offering dense vegetation, preferably bushes and trees that spread out close to the ground.

Active security measures—

- Establish security covering all likely avenues of approach into the site.
- Establish communications (wire, radio, signal, runner) with posted security to provide early warning of enemy approach.
- Select an alternate area for occupation if the original hide is compromised or found unsuitable.
- Plan for withdrawal in the event of discovery.
- Establish an alert plan with a certain percent of the personnel awake at all times.
- Organize the elements of the patrol so necessary activities can take place with a minimum amount of movement.

The size of the area physically occupied by a patrol in a hide and the number of security posts required are governed by the terrain, quantity and quality of cover and concealment, and size of the patrol.

If the situation permits, a hide can also be used as the final preparation position and/or objective rally point.

11006. IMMEDIATE ACTIONS UPON ENEMY CONTACT

A patrol may make contact with the enemy at any time. Contact may be through observation, a meeting engagement or ambush. Contact may be visual, in which the patrol sights the enemy but is not itself detected. When this is the case, the patrol leader can decide whether to make or avoid physical contact, basing his decision on the patrol's assigned mission and capability to successfully engage the enemy unit.

When a patrol's assigned mission prohibits physical contact (except that necessary to accomplish the mission), its actions are defensive in nature. Physical contact, if unavoidable, is broken as quickly as possible and the patrol, if still capable, continues its mission.

When a patrol's assigned mission permits or requires it to seek or exploit opportunities for contact (as in the case of a combat patrol), its actions are offensive in nature, immediate, and positive.

In patrolling, contacts (visual or physical) are often unexpected at very close ranges, and short in duration. Effective enemy fire often provides leaders little or no time to fully evaluate situations and issue orders. In these situations, immediate action provides a means for swiftly initiating positive offensive or defensive action, as appropriate.

Two types of physical contact with the enemy are meeting engagement and ambush. Meeting engagement is a combat action that occurs when a moving force, incompletely deployed for battle, engages an enemy at an unexpected time and place. It is an accidental meeting where neither the enemy nor the patrol expect contact and are not specifically prepared to deal with it. An ambush is a surprise attack from a concealed position.

Immediate Actions

Immediate actions are designed to provide swift and positive small unit reaction to visual or physical contact with the enemy. They are simple courses of action in which all Marines are well trained. Minimal signals or commands are required and they are developed as needed for the combat situation. The signals can, in many cases, be initiated by any member of the unit. It is not feasible to attempt to design an immediate action drill to cover every possible situation. It is better to know the immediate action drill for each of a limited number of situations that may occur during a patrol. Arm-and-hand signals associated with immediate actions—such as FREEZE, ENEMY IN SIGHT, and HASTY AMBUSH RIGHT or LEFT—are contained in FMFM 6-5 (proposed MCWP 3-11.2).

Immediate Halt

When the patrol detects the enemy but is not itself detected, the situation requires the immediate, in-place halt of the patrol. The first member visually detecting the enemy gives the silent signal for FREEZE. Every member halts in place, weapon at the ready, and remains absolutely motionless and quiet until further signals or orders are given.

Air Observation and/or Attack

These actions are designed to reduce the danger of detection from aircraft and casualties from air attack.

When an enemy or unidentified aircraft that may detect the patrol is heard or observed, the appropriate immediate action drill is FREEZE. The first member hearing or sighting an aircraft that may be a threat signals FREEZE. Every member halts in place until the patrol leader identifies the aircraft and gives further signals or orders. Members of the patrol must not look up at the aircraft as sunlight can reflect off their faces even when camouflaged.

When an aircraft detects a patrol and makes a low level attack, the immediate action drill air attack is used. The first member sighting an attacking aircraft shouts, "AIRCRAFT," followed by the direction of the incoming attack: FRONT, LEFT, REAR or RIGHT. The patrol moves quickly into line formation, well spread out, at right angles to the aircraft's direction of travel. As each member comes on line, the member hits the ground, using available cover, then positions the body perpendicular to the aircraft's direction of travel, to present the shallowest target possible (see fig. 11-2 on page 11-8). Between attacks (if the aircraft returns or if more than one aircraft attacks), patrol members seek better cover. Attacking aircraft are fired upon only on command of the patrol leader.

Meeting Engagement

Hasty Ambush. This immediate action is used to avoid contact and to prepare to initiate an unplanned ambush on the enemy. It may often be a subsequent action after the command freeze. When the signal HASTY AMBUSH is given (by the point member, patrol leader or another authorized patrol member), the entire patrol moves quickly to the right or left of the line of movement, as indicated by the signal, and takes up the best available concealed firing positions (see fig. 11-3 on page 11-9). The patrol leader initiates the ambush by opening fire and shouting, "FIRE"; thus ensuring initiation of the ambush if the weapon misfires. If the patrol is detected before this, the first member aware of detection initiates the ambush by firing and shouting. The patrol leader may decide not to initiate the ambush in order to avoid contact unless the patrol is detected. When used as an offensive

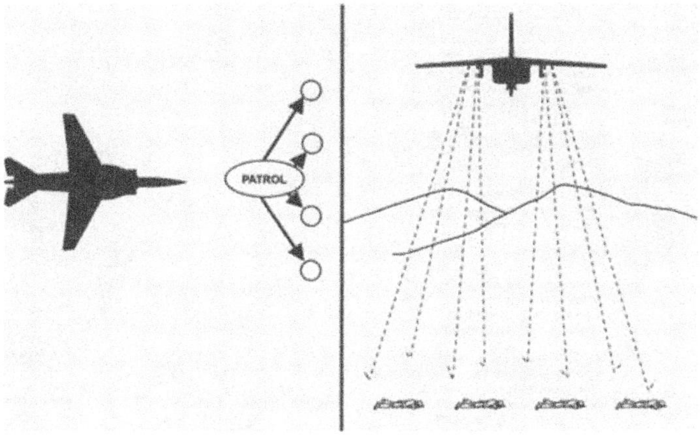

Figure 11-2. Immediate Action, Air Attack.

measure, the enemy is allowed to advance until he is in the most vulnerable position before the ambush is initiated. An alternate means for initiating the ambush is to designate an individual (for example, point or last member) to open fire when a certain portion of the enemy unit reaches or passes that member.

Immediate Assault. This immediate action drill is used defensively to make and quickly break undesired but unavoidable contact (including ambush) and offensively to decisively engage the enemy (including ambush). When used in a meeting engagement, members nearest the enemy open fire and shout, "CONTACT," followed by the direction of the incoming attack: FRONT, LEFT, REAR or RIGHT. The patrol moves swiftly into line formation and assaults (see fig. 11-4).

Defensive Measures

When used defensively, the assault is stopped if the enemy withdraws and contact is broken quickly. If the enemy stands fast, the assault is carried through the enemy positions and movement is continued until contact is broken.

Offensive Measures

When used offensively, the enemy is decisively engaged. Escapees are pursued and destroyed until orders to break contact are given by the patrol leader. If the patrol is fired upon from beyond 50 meters, the patrol must break contact as quickly as possible and continue the mission. If it engages the enemy any longer than necessary to break contact, it may put the mission in jeopardy.

Fire and Maneuver

Fire and maneuver is one means to break contact. One portion of the patrol returns the enemy fire while another portion moves by bounds away from the enemy. Each portion of the patrol covers the other by fire until contact is broken by all.

Clock System

The clock system is another means to break contact. Twelve o'clock is the direction of movement of the patrol. The patrol leader shouts a direction and a distance. For example: "TEN O'CLOCK-TWO HUNDRED," means the patrol should move in the direction of ten o'clock for 200 meters. Patrol members keep their same relative positions as they move so the original formation is not disrupted. Subordinate leaders must be alert to ensure that the members of their elements and teams receive the correct order and move as directed.

Counter Ambush

When a patrol is ambushed, the immediate action drill used depends on whether the ambush is a near ambush

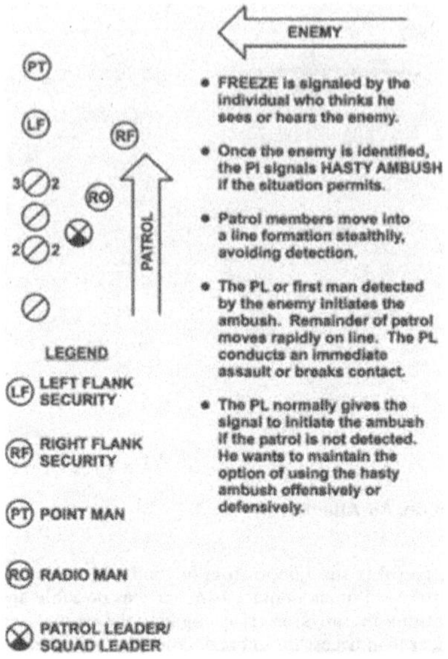

Figure 11-3. Immediate Action, Hasty Ambush.

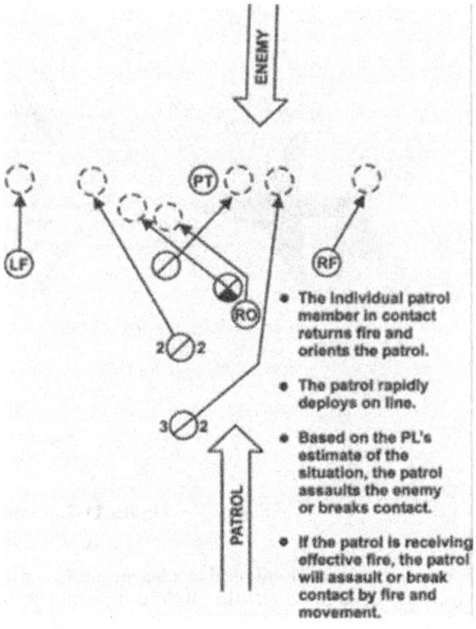

Figure 11-4. Immediate Action, Immediate Assault.

(the enemy is within 50 meters of the patrol) or a far ambush (the enemy is beyond 50 meters of the patrol). Fifty meters is considered the limit from which the ambushed patrol can effectively launch an assault against the enemy.

Near Ambush. In a near ambush, the killing zone is under very heavy, highly concentrated, close range fires. There is little time or space for members to maneu-ver or seek cover. The longer they remain in the killing zone, the greater the chance they will become casualties. Therefore, if members in the kil-ling zone are attacked by a near ambush, they immediately assault without order or signal directly into the ambush position, occupy it, and continue the assault or break contact, as directed. This action moves them out of the killing zone, prevents other elements of the ambush from firing on them without firing on their own members, and provides positions from which other actions may be taken (see fig. 11-5 on page 11-10). Members not in the killing zone maneuver against the enemy as directed. The assault continues until all patrol members are outside of the killing zone.

Far Ambush. In a far ambush, the killing zone is also under very heavy, highly concentrated fires, but from a greater range. This greater range provides members in the killing zone maneuver space and some opportunity to seek cover at a lesser risk of becoming a casualty. If attacked by a far ambush, members in the killing zone, without order or signal, immediately return fire, take the best available positions, and continue firing until directed otherwise. Members not in the killing zone maneuver against the ambush force, as directed (see fig. 11-6 on page 11-10). The assault is continued against the enemy or until the order to break contact is given.

In each situation, the success of the counter ambush employed depends on the members being well trained in quickly recognizing the distance from which an ambush is initiated and well rehearsed in the proper reaction.

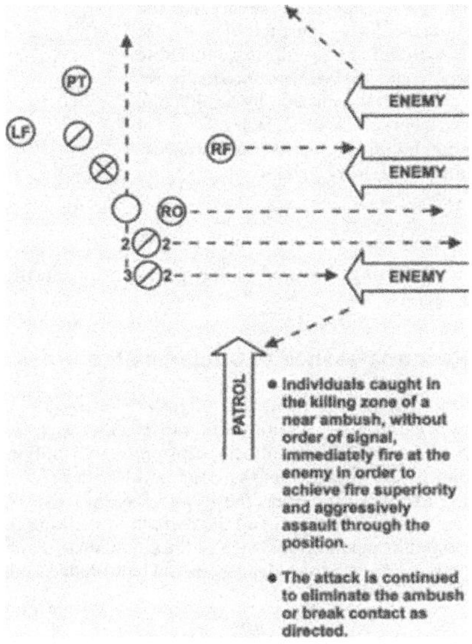

Figure 11-5. Near Ambush.

11007. PATROL LEADER'S ACTION IN A DEVELOPING SITUATION

While good patrolling depends on good planning, the patrol leader's plan must be flexible. Every combat situation develops differently than expected, and the patrol leader must be ready to quickly adapt to the situation as it develops. One of the most difficult tasks in battle is to recognize the correct moment for making a decision. Generally, it is more difficult to determine the moment for making a decision than it is to formulate the decision itself. When the situation demands, decisions must be made promptly without waiting for more information. In a developing situation, the patrol leader should use the following questions as a guide for battlefield decisionmaking:

- How has the situation changed?
- How does the change affect mission accomplishment and the immediate superior's mission?
- Must a decision be made now?
- What are the options?
- Which option best serves the mission and the unit as a whole?
- Which option offers the greatest chance of success?

11008. RETURN FROM OBJECTIVE AREA

After performing actions in the objective area, the patrol reassembles at the objective rally point. This phase of the patrol is perhaps the most difficult and dangerous. Patrol members are experiencing fatigue, emotional letdown, and wounds; they may be low on water and ammunition. Above all, the enemy was likely alerted if not in pursuit. At this point, the patrol leader must move the patrol rapidly but carefully and maintain patrol security at a high level. In returning to friendly lines, the patrol neither uses nor travels near the same route used to get to the objective area. The enemy may have that route covered. For information on reentry of friendly lines, refer to paragraph 11001.

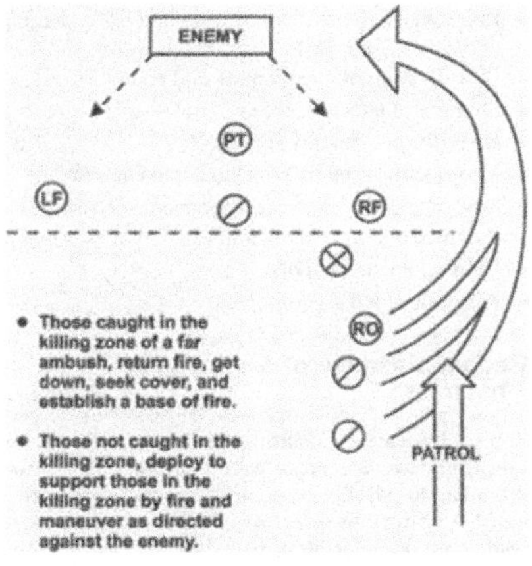

Figure 11-6. Far Ambush.

Chapter 12. Reconnaissance Patrols

The commander needs accurate, timely information about the enemy and the terrain the enemy controls to assist in making tactical decisions. Reconnaissance patrols are one of the most reliable means for obtaining this information. These patrols engage the enemy only when necessary to accomplish their mission or for protection. They generally avoid combat, accomplish the mission by stealth, and do not maintain contact with the unit that sent them out. A reconnaissance patrol is capable of carrying the search for information into the area occupied by enemy forces—usually beyond the range of vision of friendly ground observation posts—and examining objects and events at close range.

12001. General Missions

General missions for reconnaissance patrols include gathering information about location and characteristics of friendly or hostile positions and installations, terrain (routes, stream crossings), and obstacles.

12002. Specific Missions

Locate the Enemy

Reconnaissance patrols try to determine the—

- Location of enemy forces, installations, and equipment.
- Identification of enemy units and equipment.
- Strength of enemy forces.
- Disposition of enemy forces.
- Movement of enemy personnel and equipment.
- New or special types of weapons.
- Presence of mechanized units.
- Unusual enemy activity.
- Presence of NBC equipment.

Reconnaissance of Enemy Wire Obstacles

A patrol with the mission of investigating enemy-emplaced wire obstacles employs a formation providing for all-around security and takes precautions against being observed by the enemy. The patrol leader and one Marine inspect each gap and establish its location by means of compass bearings to prominent objects in the rear of enemy or friendly positions or through the use of global positioning satellite receivers.

Reconnaissance of Contaminated Areas

A patrol with the mission of investigating a contaminated area reconnoiters and marks the area's boundaries. Patrol members wear protective clothing and field protective masks. The patrol leader reports the extent of the area, the ttype of agent used, the terrain and vegetation and the method of marking the contaminated area. A sketch of the contaminated area, prepared by the patrol leader, should be included in the patrol report.

Reconnaissance of Enemy Minefields

Patrols assigned to reconnoiter enemy minefields are often composed of specially trained personnel. The type of mine is identified and the possibility of bypassing the area determined.

Reconnaissance of Terrain

The reconnaissance patrol is particularly suited for gathering information about the terrain within their area of operations. (Refer to MCRP 2-15.3B, *Reconnaissance Reports Guide*.) Commanders must know the location and condition of bridges, streams, and roads in order to make proper plans for the movement of troops. The following information should be obtained by a reconnaissance patrol assigned such missions:

- Bridges:
  - Maximum load capacity.
  - Material used in construction (wood, stone, concrete or steel).

  - Material used for abutments (dirt, wood, masonry, concrete).
  - Type and number of supporting members of the bridge.
  - Condition and dimensions of bridge flooring.
  - Presence or absence of mines and demolition charges on the bridge and at entrances/exits.
  - Terrain crossed (ravine, stream, road).
  - Primary use (rail, auto, foot).
  - Location and type of an alternate crossing if bridge is unusable.
- Streams and fords:
  - Width and depth of stream.
  - Composition of the bottom (mud, sand, gravel, rock).
  - Speed of current in miles per hour.
  - Composition and gradient of banks.
  - Surrounding terrain and vegetation cover.
- Roads:
  - Width.
  - Composition of surface.
  - Condition.
  - Road blocks.
  - Gradient/degrees of slopes.
  - Curves (location and length).
  - Pitch and culvert locations.
  - Primary use and frequency of use.

12003. TYPES OF RECONNAISSANCE

Route

Route reconnaissance is a directed effort to obtain detailed information of a specified route and all terrain from which the enemy could influence movement along that route. (Refer to MCRP 5-12C.) It is also described as a form of reconnaissance focused along a specific line of communications—such as a road, railway, or waterway—to provide new or updated information on route conditions and activities along the route. (Refer to MCRP 5-12A, *Operational Terms and Graphics*.)

A route reconnaissance normally precedes the movement of forces. It provides detailed information about a specific route and the surrounding terrain that could be used to influence movement along that route. Considerations include trafficablility, danger areas, critical points, vehicle weight and size limitations, and locations for friendly obstacle emplacement.

The objective area is normally defined by a line of departure, a route, and a limit of advance.

Area

An area reconnaissance is a directed effort to obtain detailed information concerning the terrain or enemy activity within a prescribed area such as a town, ridge line woods or other features critical to operations. (Refer to MCRP 5-12C.) An area reconnaissance could also be made of a single point, such as a bridge or installation. (Refer to MCRP 5-12A.)

Emphasis is placed on reaching the area without being detected. Enemy situations encountered en route are developed only enough to allow the reconnoitering unit to report and bypass.

Zone

A zone reconnaissance is a directed effort to obtain detailed information concerning all routes, obstacles (to include chemical or radiological contamination), terrain, and enemy forces within a zone defined by boundaries. A zone reconnaissance is normally assigned when the enemy situation is vague or when information concerning cross-country trafficability is desired. (Refer to MCRP 5-12C.)

The commander specifies specific routes or areas of interest within the zone. The zone to be reconnoitered usually is defined by a line of departure, lateral boundaries, and a limit of advance.

12004. TASK ORGANIZATION

A reconnaissance patrol is organized around the current structure of the Marine rifle squad with one or more of the squad's fire teams assigned as the reconnaissance element to reconnoiter or maintain surveillance over the objective. At least one fire team acts as a security element whose functions are to—

- Secure the objective rally point (see para. 11003).
- Give early warning of enemy approach.
- Protect the reconnaissance unit.

Reinforcing the squad is based on METT-T. If two squads are required, then the mission should be assigned to a platoon.

A company may send a reconnaissance patrol to specifically reconnoiter a given area (a riverbed or a bridge), or to maintain surveillance on a particular streambed for the next four nights. Reconnaissance patrols perform three basic functions en route to and from the objective:

- Provide control.
- Provide security while moving and in the objective area.
- Conduct reconnaissance or surveillance.

Depending on the size of the reconnaissance patrol, an element must be tasked to reconnoiter the area or zone. In the case of a surveillance mission, two teams are recommended to share the same task (so one could lay up and rest, while the other lays low and hides to maintain strict silence and no movement while surveillance is being maintained on the objective). Whatever the case, each unit that has a surveillance mission or reconnaissance mission must be thoroughly briefed as to what essential elements of information are to be collected as well as their location in the objective area.

The patrol should be organized with one or two fire teams to actually conduct the reconnaissance mission and the remaining fire team to provide security. Each fire team should be prepared to assume either mission. A small area reconnaissance patrol needs only one fire team for the assigned mission. A patrol with a wider area reconnaissance mission should use two fire teams to physically conduct the mission and one fire team for cover and/or security.

The security element for a reconnaissance patrol should be organized to cover the likely avenue of approach into the objective area, to protect the units conducting the reconnaissance, and to cover the objective rally point.

The variations of the special organization for reconnaissance patrols illustrate the patrol leader's flexibility in organizing the patrol to meet mission requirements. The patrol leader is not limited to the variations reflected herein but can choose any special organization as long as that task organization accommodates the requirement for command and control, reconnaissance, and security.

12005. SIZE OF RECONNAISSANCE PATROLS

A reconnaissance patrol should be kept to the minimum number of personnel required to accomplish the mission. A mission requiring a patrol to remain away from its unit for a considerable period of time, or one requiring a patrol to send back information by messenger, increases the size of the patrol. Reconnaissance patrols seldom exceed a squad in strength. Unit integrity should be preserved whenever possible. Intelligence personnel, interpreters, and other specialists, such as radio operators or engineers, are assigned to a patrol if the particular mission requires.

12006. RECONNAISSANCE EQUIPMENT

Patrol members are armed and equipped as necessary for accomplishing the mission. The automatic rifle in each fire team provides a degree of sustained firepower in case of enemy contact. The patrol should have at least two: pairs of binoculars, pairs of wire cutters, maps, compasses, and watches. Night observation equipment may be used. Pencils and small notebooks are carried so notes and sketches can be made. A message book with message blanks and overlay paper is mandatory.

12007. RECONNAISSANCE PATROL ACTIONS AT THE OBJECTIVE AREA

Route

The patrol leader halts and conceals the patrol near the objective area in the final preparation position. The patrol leader then conducts leader's reconnaissance to confirm the plan for positioning the security teams and employing units assigned to the reconnaissance mission. The patrol leader returns to the patrol and positions the security to provide early warning of enemy approach and secure the objective rally point. The reconnaissance unit(s) then reconnoiters the objective area (route). The reconnaissance unit may move to several positions, along or adjacent to the specific route, in order to conduct a thorough reconnaissance. After completing the reconnaissance, each reconnaissance team moves to the objective rally point and reports to the patrol leader. The patrol then returns to friendly lines and the leader makes a full report.

Area

The patrol leader halts and conceals the patrol near the objective area in the final preparation position. The patrol leader then conducts leader's reconnaissance to pinpoint the objective and confirm the plan for positioning the security teams and employing units assigned the reconnaissance mission. The patrol leader returns to the patrol and positions security to provide early warning of enemy approach and secure the objective rally point. The reconnaissance unit(s) then reconnoiters the objective area. The reconnaissance unit may move to several positions, perhaps making a circle around the objective area, in order to conduct a thorough reconnaissance. When the reconnaissance is completed, the patrol leader assembles the patrol and tells members what has been observed and heard. Other patrol members contribute observations. The patrol then returns to friendly lines and the patrol leader makes a full report.

Zone

The patrol leader halts the patrol at the final preparation position, confirms the plan, and conducts leader's reconnaissance. The patrol leader positions the security team and sends out the reconnaissance team. When the entire patrol is used to reconnoiter the zone, it provides its own security. After completing the reconnaissance, each reconnaissance team moves to the objective rally point and reports to the patrol leader. The patrol then returns to friendly lines.

CHAPTER 13. COMBAT PATROLS

Combat patrols are assigned missions that usually include engaging the enemy. They are fighting patrols. Every combat patrol has a secondary mission: gaining information about the enemy and terrain. Combat patrols are employed in both offensive and defensive combat operations and they assist the parent unit in accomplishing its mission by inflicting damage on the enemy; establishing and/or maintaining contact with friendly and enemy forces; denying the enemy access to key terrain; and probing enemy positions to determine the nature and extent of enemy presence.

13001. TASK ORGANIZATION

A combat patrol is organized around the current structure of the Marine rifle platoon. A combat patrol leader should use the unit's normal organization (fire team, squad, and platoon) in assigning functions, patrol missions, and chain of command. Combat patrols must be able to perform the following four basic functions en route to and from the objective:

- Provide control.
- Provide security.
- Provide support by fire.
- Attack or assault the objective.

A rifle platoon could task-organize as a combat patrol as follows:

- Platoon headquarters (patrol headquarters).
- First squad (security).
- Second squad (support).
- Third squad (assault).

Every combat patrol must—

- Provide a control mechanism in the form of a headquarters.
- Designate a unit (a fire team or squad) to provide security while moving en route to the objective or while at the objective. At the objective area, this unit isolates the objective area, secures the objective rally point, and covers the withdrawal.
- Designate a unit to act as support. This unit provides the base of fire in the attack or covers withdrawals or advances.
- Designate a unit(s) to conduct the attack or assault. This unit(s) engages the enemy at the objective area by fire and maneuver or movement. It also operates immediately in the objective area (searching, demolition, prisoners of war, etc.). Paragraph 9001 outlines the general organization of combat patrols. As in the case with reconnaissance patrols, the task organization of a combat patrol depends on the specific mission assigned. If any special requirements are generated because of the specific mission, the patrol is task-organized to fit the needs of the mission.

13002. EQUIPMENT

Combat patrols are armed and equipped as necessary for accomplishing the mission. In addition to binoculars, wire cutters, compasses, and other equipment generally common to all patrols, it usually carries a high proportion of automatic weapons and grenades. Communications with higher headquarters is important as success of the mission may depend on being able to call for supporting fires. Also, internal radio communications with the units and teams may be useful. However, the patrol must not be so overburdened with equipment as to impede movement or mission accomplishment.

13003. RAID PATROLS

A raid is a surprise attack on an enemy force or installation with the attacking force withdrawing after accomplishing its mission. Raids destroy or capture enemy personnel or equipment, destroy installations, or free friendly personnel who have been captured by the enemy. Patrolling techniques are used in planning and when moveing to and from the objective. (Refer to MCWP 3-41.2, *Raids*.) Surprise, firepower, and violence of action are the keys to a successful raid. Patrols achieve surprise by attacking—

- When the enemy is least prepared (e.g., during periods of poor visibility such as darkness, rain, fog, or snow).
- From an unexpected direction. (This might be accomplished by approaching through a swamp or other seemingly impassable terrain.)
- With concentration of firepower at critical points within the objective.

Planning

A successful raid requires detailed planning. The leader of a combat patrol engaged in raiding must anticipate probable situations and decide upon definite courses of action to meet them. Rehearsals are imperative.

A raid patrol conducts such missions as destroying an enemy outpost or seizing prisoners from an observation post or lightly defended position.

While preparing for the mission, the patrol leader requests fire support required for the accomplishment of the mission. If practical, artillery and mortars should be employed to isolate the objective to prevent movement of enemy reinforcements into the area.

Execution

The leader's plan must be detailed and complete. All of the considerations outlined in chapter 11 must be covered. Patrol formations must provide for ease of control and all-around security while moving to and from the objective area and provide for rapid and coordinated deployment of the various units once the objective area is reached. The leader's plan usually includes the encirclement of the hostile position—either physically or by fire—in order to isolate it during the assault.

The final simultaneous assault against the objective develops when enemy defensive fires at the objective are suppressed by either friendly fire superiority or surprise. The assault is covered by the fire of the unit assigned the function of support by fire.

Grenades, SMAWs, and demolitions are most effective for clearing bunkers.

Security units are posted to isolate the objective. The patrol leader signals them when the withdrawal begins. As a minimum, security is on each flank and to the rear (at the objective rally point).

Actions at the Objective

The patrol leader halts the patrol near the objective at the final preparation position. Security is established and the leader's reconnaissance is made with appropriate subordinate leaders. When the leaders return to the patrol, they confirm previous plans or announce any changes. Movements are arranged so all units reach their positions simultaneously. This improves the patrol's capability for decisive action, if prematurely detected by the enemy.

The teams of the security element move to positions to secure the objective rally point, give early warning of enemy approach, block avenues of approach into—and prevent enemy escape from—the objective area. As the assault element moves into position, the security element informs the patrol leader of all enemy activity, firing only if detected or on the patrol leader's order. Once the assault element commences action, the security element prevents enemy entry into or escape from the objective area. The security element covers the withdrawal of the assault element (and support element, if employed) to the objective rally point, withdrawing only upon order or prearranged signal.

As the assault element approaches the objective, it deploys early enough to permit immediate assault if detected by the enemy. Each team uses stealth while moving into proper position. On command, or if one or more of the assault element is detected and fired upon by the enemy, the support element opens fire to neutralize the objective, then ceases or shifts fire according to prearranged plans and signals. As supporting fires cease or shift, the assault element assaults the objective. Demolition, search, and other teams are protected by the assault element while they work. On order, the assault element withdraws to the objective rally point.

If a support element is employed, its leader deploys teams to provide fire support for the assault element. Each member of the support element must know the scheme of maneuver to be used by the assault element, specific targets or areas to be neutralized by fire, and the signals that will be employed to commence, shift, and cease fires. The support element withdraws on order of the patrol leader. At the objective rally point, the patrol leader quickly reorganizes the patrol and begins the return movement to friendly lines.

13004. CONTACT PATROLS

General

Contact patrols establish and/or maintain contact to the front, flanks or rear by—

- Establishing contact with an enemy force when the definite location of the force is unknown.
- Maintaining contact with enemy forces through direct and/or indirect fires, or observation.
- Avoiding decisive engagement with the enemy.

Task Organization and Equipment

Task organization and equipment depend on the known enemy situation and anticipated enemy contact. A patrol sent out to establish contact with an enemy force is organized, armed, and equipped to overcome resistance of light screening forces in order to gain contact with the main enemy force. It is not organized and equipped to engage the main enemy forces in combat. Communication is paramount; radios must be reliable over the entire distance covered.

Actions at the Objective

The patrol leader selects a series of objectives. Once an objective is reached, the patrol leader initiates a planned set of actions in order to establish and maintain contact with the enemy. These plans and actions are guided by the missions to establish or maintain contact—not to engage in decisive combat. Contact with the enemy is maintained for the purposes of surveillance, applying pressure, and preventing seizure of the initiative. If the contact patrol becomes decisively engaged with the enemy, many of the tasks originally assigned to the patrol cannot be accomplished, since the enemy has seized the initiative and friendly forces are not forced to react.

13005. AMBUSH PATROLS

General

An ambush is a surprise attack from a concealed position upon a moving or temporarily halted target. It is one of the oldest and most effective types of military actions. Ambush patrols conduct ambushes of enemy patrols, resupply columns, and convoys. The intent of an ambush is to place the enemy in a dilemma where staying in the kill zone or attempting to move out of it prove equally lethal. The ambush may include an assault to close with and decisively engage the enemy, or the attack may be by fire only.

Purpose of Ambushes

Ambushes are executed for the general purpose of reducing the enemy's over-all combat effectiveness and for the specific purpose of destroying its units. The cumulative effect of many small ambushes on enemy units lowers enemy troop morale and harasses the enemy force as a whole.

Destruction is the primary purpose of an ambush because loss of men killed or captured, and loss of equipment and supplies destroyed or captured, reduces the overall combat effectiveness of the enemy.

Harassment, though less apparent than physical damage, is a secondary purpose of ambushes. Frequent ambushes force the enemy to divert men from other missions to guard convoys, troop movements, and carrying parties. When enemy patrols fail to accomplish their mission because they are ambushed, the enemy is deprived of valuable information. A series of successful ambushes causes the enemy to be less aggressive and more defensive minded. His men become apprehensive, overly cautious, reluctant to go on patrols, seek to avoid night operations, are more subject to confusion and panic if ambushed, and in general, decline in effectiveness.

Classification of Ambushes

A *deliberate ambush* is one in which prior information about the enemy permits detailed planning before the patrol departs for the ambush site. Information needed to plan a deliberate ambush includes the size, composition, and organization of the force to be ambushed; how the force operates; and the time it will pass certain points or areas. A deliberate ambush may be planned for such targets as—

- Any force if sufficient prior information is known.
- Enemy patrols that establish patterns by frequent use of the same routes or habitually depart and re-enter their own areas at the same point.
- Logistic columns.
- Troop movements.

An *ambush of opportunity* is conducted when available information does not permit detailed planning before the patrol departs. This is the type of ambush that an infantry unit normally conducts. An ambush of opportunity should not be confused with a hasty ambush. An ambush of opportunity is a planned ambush; a hasty ambush is an immediate action. In planning for an ambush of opportunity, the patrol must be prepared to execute any of several courses of action based on the types of targets that may be ambushed and must rehearse prior to departure. The course of action taken is determined when the opportunity for ambush arises.

The patrol leader may be directed to reconnoiter an area for a suitable ambush site, set up at the site selected, and execute an ambush against the first profitable target that appears.

The patrol may depart just after dark, move to a specific point, observe until a designated time, ambush the first profitable target after that time, and return before daylight.

A *hasty ambush* is an immediate action where the patrol makes visual contact with an enemy force and has time to establish an ambush without being detected. The actions for a hasty ambush must be well rehearsed and accomplished through the use of hand and arm signals given from the patrol leader.

Types of Ambushes

There are two types of ambushes: point and area. The *point ambush* is one where forces are deployed to attack along a single killing zone. The *area ambush* is one where forces are deployed as multiple related point ambushes.

A point ambush, whether independent or part of an area ambush, is positioned along the enemy's expected route of approach. Formation of the forces conducting the ambush is important because, to a great extent, the formation determines whether a point ambush is able to deliver the heavy volume of highly concentrated fire necessary to isolate, trap, and destroy the enemy.

The ambush formation to be used is determined by careful consideration of possible formations and the advantages and disadvantages of each in relation to—

- Terrain.
- Visibility.
- Forces.
- Weapons and equipment.
- Ease or difficulty of control.
- Target to be attacked.
- Combat situation.

For a detailed discussion of ambush formation, see appendix D.

Ambush Operation Terms

The *ambush site* is the location where an ambush is established.

The *killing zone* is that portion of an ambush site where fires are concentrated to trap, isolate, and destroy the target. On little-traveled roads, an obstacle placed in a defile, in the woods, on a bridge or on a steep upgrade can be used effectively to force vehicles to halt, and thus render the occupants vulnerable to attack. Antitank mines may be emplaced and the occupants of the wrecked vehicle killed or captured while still dazed by the explosion.

A *near ambush* is a point ambush where the attacking force is located within reasonable assaulting distance of the killing zone (50 meters is a guide). A near ambush is most often conducted in close terrain, such as a jungle or heavy woods.

A *far ambush* is a point ambush where the attack force is located beyond reasonable assaulting distance of the killing zone (beyond 50 meters is a guide). A far ambush may be more appropriate in open terrain offering good fields of fire or when the target will be attacked by fire only.

Factors for a Successful Ambush Patrol

There are many factors that give the ambush its best chance of success. The ideal situation would be to position the ambush on favorable terrain and have detailed planning completed beforehand.

The patrol leader of an ambush looks for the most favorable terrain in which the enemy is canalized between two obstacles with limited opportunity to attack or escape. Suitable areas include defiles, small clearings, bends in trails, and steep grades. Dense undergrowth adjacent to the ambush site permits observation from concealed positions. The ambush patrol should have maximum cover and concealment, not only for the firing positions, but for the routes of withdrawal. The enemy should be in an area offering as little protection from fire as possible. Favorable fields of fire include stretches of road, trail or open ground of at least 100 meters for machine guns and 15 meters for rifle fire and grenades. The ambush site can be improved by constructing obstacles—such as felled trees, wire, land mines, or booby traps—to impede the enemy.

Planning

A deliberate ambush or an ambush of opportunity requires thorough planning.

A deliberate ambush plan is based on extensive knowledge of the enemy and terrain, and is planned and rehearsed in great detail. A physical reconnaissance of the ambush site is made during the preparation phase and information gained is incorporated into the plan. All likely immediate actions of the enemy when ambushed are examined. Planned counteractions are developed and rehearsed.

In planning an ambush of opportunity, any available information on the enemy and terrain is used. A tentative plan for the ambush that incorporates all anticipated actions is developed and rehearsed. However, the bulk of planning is done concurrently during the patrol leader's reconnaissance of the prospective ambush site. In a rapidly developing situation, hasty ambush immediate action is employed.

The route and ambush site considerations apply to both deliberate ambushes and ambushes of opportunity. A primary route that allows the patrol to enter the ambush site from the rear is planned. Entering the prospective killing zone is avoided. If the killing zone must be entered to place mines or explosives, care is taken to remove any tracks and signs that might alert the enemy and compromise the ambush. If mines or explosives are to be placed on the far side of the ambush site, or if the appearance of the site from the enemy's viewpoint is to be checked, a wide detour is made around the killing zone. Care is taken to remove any tracks that might reveal the ambush. An alternate route from the ambush site to the objective rally point, as in other patrols, is planned.

Maps and aerial photographs are used to carefully analyze the terrain. When possible, an on-the-ground reconnaissance of the ambush site is made prior to occupation. Obvious ambush sites are avoided as the element of surprise is even more difficult to achieve in these areas. An ambush site must provide for—

- Favorable fields of fire.
- Occupation and preparation of concealed positions.
- Canalization of the target into the killing zone. (An ideal killing zone restricts the enemy on all sides, confining him to an area where he can be quickly and completely destroyed. Natural obstacles, such as cliffs, streams, embankments, or steep grades, are used whenever possible to force vehicles to slow down. Man-made obstacles, such as barbed wire, mines, and craters in the roads, are used to supplement natural obstacles.)
- Covered routes of withdrawal that enable the ambush force to break contact.
- Avoidance of enemy pursuit by fire.

Occupation of Ambush Site

The surrounding area is searched for enemy patrols prior to occupation of the ambush site. Ambush formations are used to physically deploy the patrol in a manner to inflict maximum destruction upon the enemy and to provide maximum security to the patrol. Ambush formations are contained in appendix D.

Positions

The patrol is moved into the ambush site from the objective rally point. Security is positioned first to prevent surprise while the ambush is being established. Automatic weapons are then positioned so each can fire along the entire killing zone. If this is not possible, overlapping sectors of fire are provided to cover the entire killing zone. The patrol leader then selects his position, located so he can tell when to initiate the ambush. Riflemen and grenadiers are positioned and sectors of fire are assigned to cover any dead space left by the automatic weapons. The patrol leader sets a time by which positions are to be prepared. Patrol members clear fields of fire and prepare positions in that order, with attention to camouflage for both.

Suitable Objective Rally Point

An easily located objective rally point is selected and made known to all patrol members. The objective rally point is located far enough from the ambush site so that it will not be overrun if the enemy assaults the ambush. Routes of withdrawal to the objective rally point are reconnoitered. Situation permitting, each man walks the route he is to use and picks out checkpoints. When the ambush is to be executed at night, each man must be able to follow his route in the dark. After the ambush has been executed, and the search of the killing zone completed, the patrol is withdrawn quickly but quietly, on signal, to the objective rally point where it reorganizes for the return march. If the ambush was not successful and the patrol is pursued, withdrawal may be by bounds. The last group may arm mines, previously placed along the withdrawal route, to further delay pursuit.

Local Security

Security must be maintained. Security elements do not usually participate in the initial attack, but protect the rear and flanks, and cover the withdrawal.

Patience

The Marines of the ambush force must control themselves so that the ambush is not compromised. Patience and self-discipline are exercised by remaining still and quiet while waiting for the target to appear, particularly if the patrol occupies the ambush site well ahead of the arrival of the enemy. Patience is necessary so as not to alert the enemy to the presence of the ambush.

Surprise

Surprise must be achieved, or the attack is not an ambush. If complete surprise cannot be achieved, it must be so nearly complete that the target is not aware of the ambush until too late for effective reaction. Surprise is achieved by careful planning, preparation, and execution so that targets are attacked when, where, and in a way for which they are least prepared.

Coordinated Fires

Properly timed and delivered fires contribute heavily to the achievement of surprise, as well as to destruction of the enemy. The lifting or shifting of fires must be equally precise; otherwise, the assault is delayed and the enemy has an opportunity to recover and react. All weapons, mines, and demolitions are positioned and all fires, including those of available artillery and mortars, are coordinated to achieve the following results:

- Isolation of the killing zone to prevent the enemy's escape or reinforcement.
- Surprise delivery of a large volume of highly concentrated fires into the killing zone.

Control

Close control of the patrol is maintained during movement to, occupation of, and withdrawal from the ambush site. This is best achieved through rehearsals and establishment and maintenance of good communications. When the enemy approaches, the temptation to open fire before the signal is given is resisted. The patrol leader must effectively control all elements of the ambush force. Control is most critical at the time the enemy approaches the killing zone. Control measures must provide for—

- Early warning of enemy approach.
- Fire control. Withhold fire until the enemy has moved into the killing zone, then open fire at the proper time.
- Initiation of appropriate action, if the ambush is prematurely detected. Individual patrol members must be prepared to react if detected by the enemy prior to the initiation of the ambush.
- Timely and orderly withdrawal of the ambush force from the ambush site and movement to the objective rally point.

It is important to remember that an ambush patrol should have four distinct signals: one to open fire (with an alternate signal to open fire to be used at the same time as the primary); a signal to cease fire or shift fire; a signal to assault or search the killing zone; and a signal to withdraw. The signal to open fire should meet two criteria: first, it should be the firing of a weapon that will kill the enemy; secondly, it should be a weapon reliable in any weather condition. A good primary signal is a Claymore mine, and an alternate signal would be a closed bolt weapon (M16A2). Open bolt weapons (M240G, M249) should not be relied upon to initiate an ambush.

Execution of an Ambush

The manner in which the patrol executes an ambush depends primarily on whether the ambush's purpose is harassment or destruction. To a lesser degree, the execution of the ambush is determined by whether the ambush is deliberate or an ambush of opportunity.

When the primary purpose is harassment, the patrol seals off the area with security teams to prevent enemy reinforcement and escape. Maximum damage is inflicted with demolitions and automatic weapons fire. The patrol delivers a very heavy volume of fire for a short time and withdraws quickly and quietly. The patrol avoids being seen by the enemy.

When the primary purpose of the patrol is destruction, the area is sealed off with security units. Maximum damage is inflicted with demolitions, antitank weapons, and automatic weapons fire from the support team or element. When these fires cease or shift, an assault is launched into the killing zone with heavy fire and violence to complete destruction. The assault unit provides security, while designated teams search and/or capture personnel and destroy vehicles and equipment. On the patrol leader's command, or by prearranged signal, all units withdraw to the objective rally point and move out quickly.

When the patrol's primary purpose is to obtain supplies or capture equipment, security units seal off the area. Demolitions and weapons are used to disable vehicles. The assault unit must use care to ensure its fire does not damage the desired supplies or equipment. Designated teams secure the desired items; other teams then destroy enemy vehicles and equipment.

The most successful ambush is one where the attacker is deployed and concealed in such a way that the enemy will unknowingly be surrounded by fire. The usual method is for the attackers to deploy themselves along a trail or route the enemy will travel. The enemy is permitted to pass by the center of the attacker's force so that the attack can be made from the front, flank, and/or rear. One or two men are posted well forward and to the rear along the route to prevent any enemy from escaping. All fires should be delivered simultaneously on a prearranged signal.

An effective method of luring the enemy is for an ambushing patrol to cut communication or electrical wire. The patrol then deploys and ambushes the line repair crew when it arrives. Since the line crew may be protected by riflemen, the attackers must be careful to engage the entire party.

Vehicles and foot personnel moving on well-established transportation routes can sometimes be captured by altering or moving directional signs so as to divert the enemy into an area where he can be more readily attacked. The attack can best be accomplished at an obstacle, such as a stream or gully, that forces the enemy to stop or slow down.

After the enemy has been ambushed and destroyed, the unit quickly withdraws over a prearranged route to the objective rally point. Speed is very important, since the noise of the ambush could alert other nearby enemy units.

13006. SECURITY PATROLS

General

Security patrols are assigned missions that may or may not require them to engage the enemy. They are used in proximity to defensive positions, on the flanks of advancing units or in rear areas. Purposes of security patrols are to detect infiltration by the enemy, destroy infiltrators, and protect against surprise and ambush.

In any situation where there is a threat of attack, such as a rear area threatened by guerrillas or a facility that is under threat of a terrorist attack, all Marines, not just the infantry, must know how to conduct a security patrol.

In just the offensive operations, infantry units provide security patrols to screen their flanks, areas, and routes. Whereas, in defensive operations, security patrols are used to prevent the enemy from infiltrating an area, detect and destroy infiltrators, and prevent surprise attacks. In rear areas, particularly when there is guerrilla or terrorist threat, the requirement to conduct security patrols increases for all Marine air-ground task force (MAGTF) units ashore, particularly aviation and combat service support units.

Task Organization and Equipment

Generally, a Marine rifle squad or similar sized organization is considered ideal for security patrols. Communications are important to higher headquarters so that they receive information from the patrol; and communications are important to the patrol to request fire support, etc. The radio the patrol carries must have the range necessary for higher headquarters to be able to receive transmissions from anywhere along the patrol route, and the patrol must have a secondary means of reporting (i.e., flare signals upon contact).

Patrol Procedures

All of the procedures presented in previous sections are to be used in security patrols.

Patrol Planning

Security patrol planning includes—

- Rehearsing prior to departing friendly lines.
- Maintaining communications.
- Support by organic weapons.
- Reinforced if necessary.
- Using varied routes and never establishing a routine pattern.
- Staying within proximity of friendly units.

Patrol Techniques

Within rear areas, an irregular pattern of patrol is established and changed daily. Outside of friendly lines it would be prudent to establish a definite preplanned route for the patrol, of which all adjacent units know the route. The parent unit commander establishes frequent checkpoints for control. If checkpoints are designated, the patrol leader treats them as individual objectives to be searched and cleared.

The patrol has a definite plan as to what to do if contact with enemy is made, how to break contact, how to defend itself, and how to call for supporting fires. It is imperative that patrol members know what to do if they become split or separated; i.e., location of rally points and how to be recovered.

13007. URBAN PATROLS

General

As national strategy continues to focus on regional vice global conflicts, the Marine Corps will continue to conduct urban operations in various operational environments. Cities and towns are often the center of economic and political power and are therefore extremely vulnerable to urban insurgent activities and violence. The lessons learned from recent operations in Somalia, as well as experiences gained by British forces in Northern Ireland, provide the foundations of urban patrolling.

Enemy Forces

Enemy forces in urbanized areas range from organized military forces to low intensity engagements with insurgents, such as terrorists or local gangs. For Marine Corps doctrine and tactics, techniques, and procedures (TTP) for dealing with the higher intensity threat in urbanized areas, refer to MCWP 3-35.3, *Military Operations on Urbanized Terrain.*

Common Threat Tactics

Cities provide cover and concealment for both friendly forces and enemy forces. However, enemy forces generally find active support only in certain areas of the town or city. The urban-based insurgent or terrorist usually lives in a friendly community or in one where the people are too frightened to withhold support or inform anyone about the situation. The insurgent or terrorist normally maintains close contact with leaders and others friendly to the cause. The enemy will often have efficient communication and intelligence systems, sometimes involving women and children to provide cover for its activities.

The urban insurgent normally cannot, like his rural counterpart, establish bases and recruit large military units. He is generally an individual or a member of a relatively small group. He relies on the cover afforded by the city's populace and terrorizing them to coerce loyalty or support.

Urbanized areas tend to give the insurgent and/or terrorist many opportunities to initiate action and gain advantage. The normal presence of large numbers of people in cities provides the foe an opportunity to mass crowds quickly and manipulate demonstrations easily. The presence of women and children during mass demonstrations may restrict the courses of action available to friendly forces. Major incidents stemming from overreaction or excessive use of force by friendly forces may provide the insurgent with propaganda material. Publicity is easily gained in an urbanized area because major incidents can't be completely concealed from the local population. Insurgent successes can be exploited to discredit the ability of host nation police, friendly forces, and the civil government, and gain recognition for the insurgents' cause.

The urban insurgent or terrorist can usually be expected to operate more boldly than his rural counterpart. This is reflected in the enemy's tactics. A single sniper or bomber may be the norm in the urbanized area, whereas the rural threat is generally the more conventional ambush. In urbanized areas, explosive devices can be easily emplaced and used effectively against large groups or select individuals. Enemy forces may be expected to employ the following tactics in urbanized areas:

- Using local communications, such as radio and newspapers, for propaganda purposes.
- Disrupting industry and public services through strikes and sabotage.
- Generating widespread disturbances designed to stretch the resources of the security force.
- Creating incidents or massing crowds in order to lure the patrol or reaction force into a trap.
- Provoking security forces in the hope that they may react improperly, therefore discrediting the security force by means of propaganda.
- Sniping at roadblocks, outposts, sentries, and patrols.
- Attacking friendly bases with rockets and mortars.
- Planting explosive devices, either against specific targets or indiscriminately, to cause confusion and destruction, and lower public morale and confidence.
- Using ambush patrols.
- Firing on friendly helicopters.

Principles of Urban Patrolling

Patrolling in an urban environment often presents conditions considerably different and often more complex than those encountered in rural and less inhabited areas. While the principles of patrolling are still relevant in an urban situation, the nature of urban patrolling has led to the development of six specific urban patrolling principles. They are—

1. **Depth.** The restrictive, canalizing nature of urbanized terrain usually limits a patrol's ability to disperse laterally. To prevent the patrol from bunching up, patrols normally maintain dispersion along the length of a patrol formation.

2. **Mutual Support.** The positioning of units in-depth within the patrol enables one unit to cover another unit's movement and facilitates immediate action during various situations. Aircraft, vehicles, and snipers also provide good mutual support.

3. **Deception and Pattern Avoidance.** Deception and pattern avoidance are normally a planning consideration of the headquarters directing the entire patrolling plan. By varying patrol routes, durations, and departure times, hostile actions commonly used against urban patrols—such as enemy ambushes, and roadblocks—are more difficult to plan and may be preempted.

4. **Intra-patrol Communication.** Elements of an urban patrol must have the means to communicate with each other. Ideally, each element will possess a radio enabling it to remain in continuous communication that facilitates rapid response and reporting to higher headquarters, reaction force coordination, and coordination of actions with other patrols or fire support agencies.

5. **Establishment of a Reaction Force.** The nature of urbanized terrain (its compartmentalization) makes urban patrols more vulnerable to a wider range of hostile actions. The requirement for immediate, coordinated reinforcement of a patrol is best satisfied by employment of an established reaction force. The reaction force requires superior mobility (relative to the enemy's) and fire support to be effective for this task.

6. **Three-Dimensional Threat**. Patrolling in an urban environment requires constant attention to its three-dimensional aspect; hostile actions can originate from rooftops, streets, subsurface levels or combinations of all levels at once.

Classification of Urban Patrols

Mission

The vast majority of urban patrols are overt in nature, with their presence readily apparent to the local populace. Most urban patrols are combat vice reconnaissance patrols. The vast majority of urban patrols are security patrols. However, units may be assigned secondary tasks of reconnoitering specific or general areas along the patrol routes.

Raids normally involve a swift penetration of an objective to secure information, confuse the enemy or destroy installations. Raids include a planned withdrawal upon completion of the assigned mission.

Movement

Means of movement are as follows:

- Dismounted: movement on foot.
- Vehicular: movement by motorized, mechanized or armored vehicle.
- Helicopterborne: movement by helicopter, however, helicopterborne patrols will usually involve dismounted or vehicular movement after patrol insertion.
- Combination: movement using a combination of methods.

Task-Organization

Units task-organize to meet the specific requirements of the mission and situation. The Marine rifle squad is ideally suited for urban patrolling and can easily integrate attached specialists required by specific missions. Specialists who may accompany urban patrols include—

- Interrogator-translator team (ITT) and counterintelligence team (CIT) Marines.
- Explosive ordnance disposal (EOD) personnel.
- Members of host nation or allied military forces.
- Interpreters.
- Local community leaders.
- Local law enforcement officers.
- Public affairs personnel escorting media representatives.

Dismounted Patrol Organization

Patrols should maintain unit integrity (fire team and/or squad) in organizing elements for an urban patrol. Similar to reconnaissance patrols, the vulnerability of urban patrols necessitates that all elements must provide for their own security in addition to the combat aspect of the mission. Combined assault and security teams are an effective method to organize for all-around security.

Urban Patrol Base Operations

Urban patrols may operate from an established patrol base that may be located within the unit's assigned area of operations or an area designated for a patrol. The patrol base should be located in a building used exclusively for this purpose. Patrol bases may also be located within a larger site that houses other agencies, such as a higher headquarters' command echelon.

If located within a larger site, the patrol base will be included in the overall facility security plan. If isolated, the patrol base must consider the following security factors:

- External security:
  - Barrier plan.
  - Sentry posts.
  - Local security patrols.
- Internal security:
  - Covered positions for all Marines.
  - Contingency plan for hostile actions against patrol base.

The Reaction Force

The high probability of an urban patrol becoming involved in a hostile or volatile incident requires the establishment of a dedicated reaction force for rapid reinforcement, support or extraction of the patrol. Ideally, the reaction force is—

- Large enough and task-organized in a manner that it can meet and quickly defeat the expected threat. Reaction forces are normally tiered with a lead element (normally one-third of the unit's size) and a main body force (the other remaining two-thirds of the force). Reaction force response times routinely are determined in advance by higher headquarters.
- Ready to respond immediately.
- Motorized or mechanized and supported by close air support and other fire support.
- Familiar with the area of operations.
- Briefed on the patrol's plans and monitors the status of patrols in progress.
- Task-organized to be multimission capable.
- Able to communicate with the higher headquarters, fire support assets, patrol base, and the patrol.
- Controlled by higher headquarters, once employed.

Patrol Preparation

Planning

Higher headquarters will—

- Designate the area for patrol.
- Provide intelligence briefs and updates.
- Ensure liaison with allied forces and the civilian populace.
- Provide special equipment and personnel required for the mission (scout snipers, public affairs officer, interpreters, etc.)
- Provide urban maps, photos, terrain models as required.
- Consider deception and pattern avoidance when issuing mission.
- Prescribe rules of engagement (ROE).

Intelligence Brief

An intelligence brief is conducted by the S-2 officer or representative prior to a patrol conducting its mission. The brief adresses the situation relevant to the specific patrol (e.g., routes, areas, updated enemy situations).

Coordination

Higher headquarters will effect liaison with adjacent and allied forces, as well as civil authorities and other agencies, having a possible effect on the patrol. The patrolling unit generally follows the same procedures as those used during patrol planning and execution in a jungle or forest environment.

The Urban Patrol Order or Warning Order

The Urban Patrol or Warning Orders use the same format and considerations as noted in this publication for patrolling. They rely heavily on a detailed terrain model, photographs, and subterranean construction to ensure complete understanding of the plan.

Rehearsals

The limited size of the patrol base usually precludes the need for full-scale rehearsals. Immediate action drills, such as crossing danger areas, are rehearsed in as much detail as possible, despite the limited available space.

Inspections

Initial and final inspections are conducted in the same manner as other patrols. Attached personnel must be fully integrated into the patrol and familiar with the plan and unit standing operating procedures (SOPs).

Conducting an Urban Patrol

Movement

Individual and unit movement considerations are generally the same as those for other patrols. However, urban environments require consideration of additional factors. Because of these factors, an urban patrol leader should—

- Ensure that each movement within a patrol takes place under the observation or cover of another individual or element of the patrol.
- Know where cover can be taken in the event of a hostile incident or action.
- Be prepared for contact with civilians, especially children, during the patrol and be aware that they may intentionally attempt to distract patrol members.
- Expect the presence of vehicles (both moving and stationary) along the patrol route.
- Expect members of the patrol to be approached by dogs and what action to take if threatened.

Patrol Formations

Squad-sized Patrols. The need for immediate fire power outweighs the dangers of becoming canalized. In contrast to other types of patrols, the headquarters element of an urban patrol will normally locate at the lead of the patrol column. This allows the patrol leader greater flexibility through control of two combined assault and security (A&S) teams. The leaders of these elements tactically stagger their members on each side of the road (see fig. 13-1 on page 13-12).

A&S teams follow in trace of the headquarters element and maintain unit integrity on separate sides of the street. One unit will remain slightly to the rear to create a staggered interval between Marines on either side of the street. This allows A&S teams to take lateral routes in support of headquarters element without having to cross a street to do so.

Platoon-sized Patrols. Squads will generally travel abreast of each other, moving along parallel routes. The interval between squad-sized units and/or teams is situation-dependent, but is usually between 100 and 150 meters (roughly two city blocks; this often prevents visual contact between the units). The intent is to create less of a target to an aggressor, yet still allow the patrol to quickly react to an incident. Individuals within units or teams will move in a staggered column as in a squad-sized patrol. See figure 13-2, which is located on page 13-13).

Night

Night patrols will generally be at least squad-sized and will generally use the same formation as that for day patrols. At night, it may be necessary to close distances between individuals or elements to maintain control. Consideration should be given to the use of night vision devices and thermal weapons sights.

Navigation, Control, and Security Measures

The patrol leader is ultimately responsible for the navigation. The headquarters element normally functions as the base unit during movement. The designated navigator is normally assigned from within the headquarters element. City maps are often inaccurate or outdated; however, when used with aerial photographs and other navigational aids, they can be effective for urban navigation.

Checkpoints and phase lines should be related to major streets (alleys, buildings, bridges) for easy identification. Arrival at checkpoints and crossing of phase

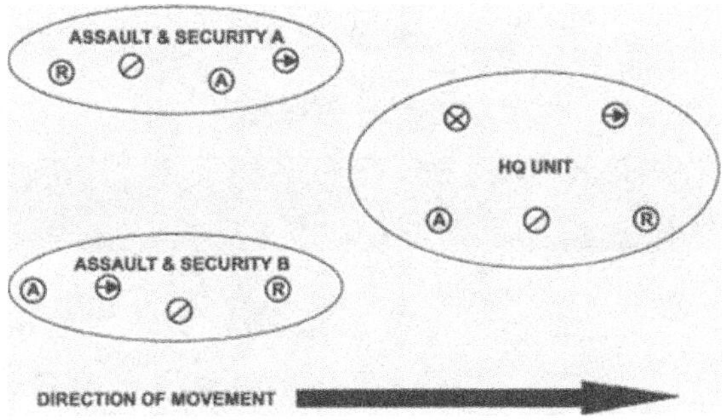

Figure 13-1. Squad-sized Dismounted Urban Patrol Formation.

lines should be relayed to higher headquarters using established brevity or codewords. A detailed patrol overlay is issued to both the reaction force and higher headquarters to keep them advised of routes and control measures used. A terrain model should be maintained at higher headquarters to aid in supervision and control of the patrol.

The use of camouflage should be limited to avoid frightening and confusing the local populace. The often overt nature of urban patrols may negate the need for camouflage. Patrols are sometimes deployed to show force presence and usually move on the urban street in plain view.

During daylight, patrols will routinely vary their rate of movement ranging from short halts to brief periods of double-timing. The British term for this urban patrolling technique is *hard-targeting*, meaning it makes the patrol harder for an enemy to target. Altering the rate of movement is intended to frustrate the enemy's ability to coordinate an attack or ambush against a targeted patrol.

Patrols should use short security halts, with Marines taking up mutually supporting firing positions. Marines must always work in pairs, ensuring mutual support. The last Marine in the element will provide rear security, but stays in his buddy's sight.

Individual Tasks

Individuals may be assigned collateral tasks performed throughout the patrol that may increase the patrol's situational awareness. Individual tasks may include—

- Vehicle spotter: looks for suspicious or known insurgent vehicles.
- Personnel spotter: observes and attempts to recognize previously identified enemy in crowds.
- Talker: attempts to gain information from casual conversations with the local populace. (Talkers are usually subordinate leaders or Marines with foreign language skills.)
- Searcher: conducts physical searches of vehicles and personnel while other patrol members provide cover and security.
- Marksman: engages point targets when the tactical situation does not permit massed or high volume fire. Other patrol members provide security to cover the marksman's engagement.

Departure of Friendly Lines

Urban patrols must vary their departure times to prevent being ambushed while exiting the patrol base.

Individual elements will usually depart exits at staggered times and at different movement rates, especially where sentries cannot provide cover. (This technique is used when the environment adjacent to the patrol base is dangerous.)

Once an element has exited friendly lines, a short halt is conducted in a predetermined, covered initial rally point 50 to 100 meters from the base. This ensures all elements are in position before the patrol continues.

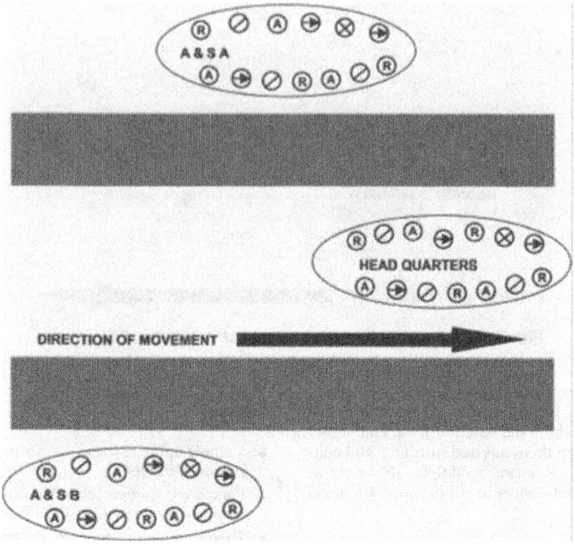

Figure 13-2. Platoon-sized Dismounted Urban Patrol Formations.

Exit points, routes from the base, departure techniques, and locations of IRPs should be varied constantly to avoid setting a pattern. This pattern avoidance may also include using vehicles to insert patrols away from the urban patrol base and employing empty vehicles as part of a deception plan.

Danger Areas

Urban patrols may encounter hundreds of danger areas during a single patrol. The three-dimensional threat requires keen situation awareness by every patrol member. Many danger areas can be dealt with simply by avoidance, while others require an adjustment of patrol formation, movement rate, etc. In the urban environment, places to be treated as danger areas are points that pose a major threat to the patrol, such as local political and religious headquarters, weapons containment areas, roads and routes that canalize movement and direct fire, and any area with a history of repeated contact.

Near and far side rally points are designated and briefed during the issuance of the patrol order. Squad-sized formations may use the A&S teams to provide flank security for the headquarters element and for each other. The headquarters element identifies the danger area and takes up a position on the near side of intersection. Individuals provide all-around security (see fig. 13-3 on page 13-14).

Two Marines (one from each A&S team) are designated to move through the headquarters element and establish respective firing positions on the near side of the danger area covering the patrol's near side flanks. They are followed by a second pair (again, one Marine from each A&S team) that moves across to the far side of the danger area and establishes respective firing positions covering the patrol's far side flanks (see fig. 13-4 on page 13-14).

Once near and far side flank security is established, the headquarters element moves across to the far side of danger area (see fig. 13-5 on page 13-15).

The remaining A&S team members then cross the danger area and join the headquarters element on the far side (see fig. 13-6 on page 13-15).

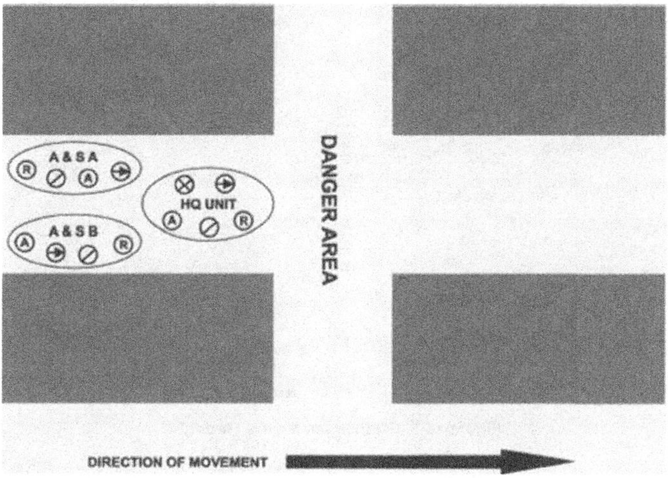

Figure 13-3. Approaching Danger Area.

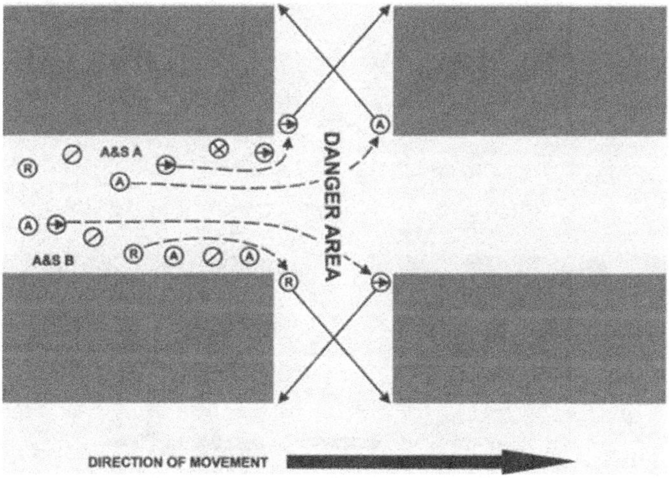

Figure 13-4. Securing Flanks.

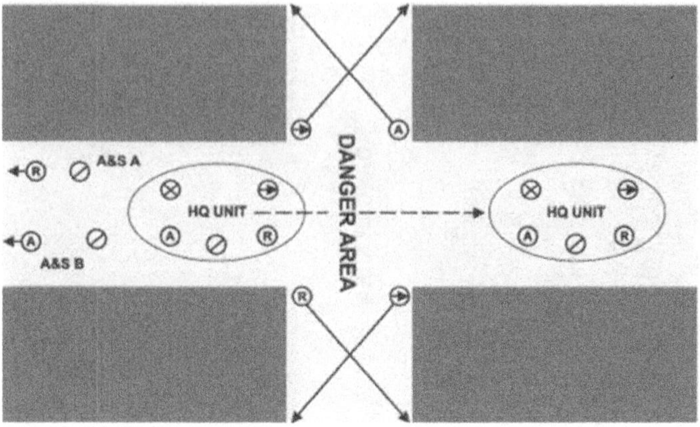

Figure 13-5. HQ Element Moves Through.

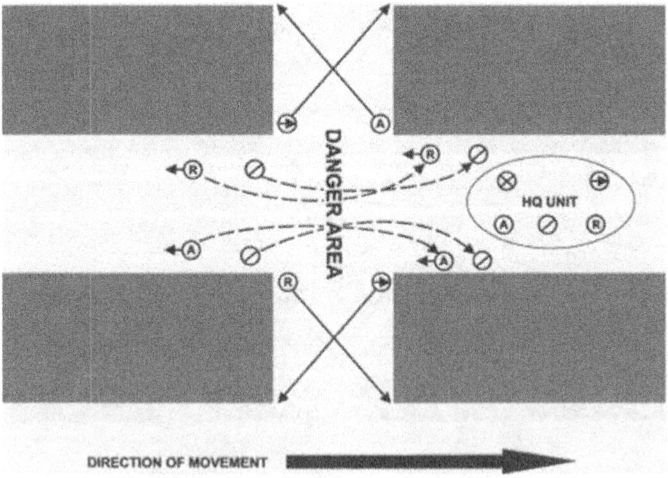

Figure 13-6. A&S Elements Move Through.

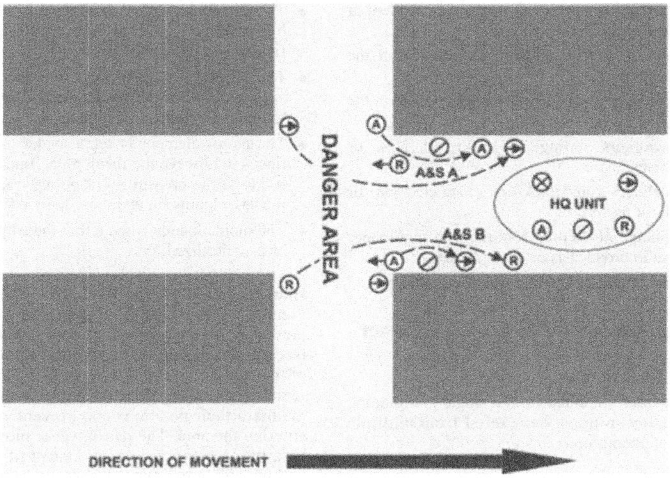

Figure 13-7. Patrol Resumes Movement Along Intended Route.

Once the trail A&S team members cross the danger area, they take up rear security and cover the movements of the near and far side flank security teams as they return to their positions in the patrol formation. The near side security team should collapse back first, followed by their far side counterparts. The patrol then resumes its advance away from the danger area (see fig. 13-7).

Interaction with Local Populace

Urban patrols must interact with the local populace. Patrols are at first a novelty to the civilians but can quickly become an unwelcome intrusion. The movement of the patrol must be fast enough to prevent the enemy from massing their fires upon it, but deliberate enough to ensure adequate security and mutual support. Patrol members must realize that they are usually the only Marines the local populace will encounter and that an inappropriate gesture, comment or act could lead to the deterioration of rapport between U.S. forces and the general population. Marines must remember that the vast majority of the individuals with whom they come in contact will be noncombatants attempting to survive in trying political, economic, and social situations.

Hostile incidents often seem to occur spontaneously, but there are usually indications that can alert Marines to imminent danger. The most obvious are the sudden alteration of normal routines, patterns, and attitudes of the local populace or other unusual activity. Some examples include—

- Observers on rooftops, in windows, etc., who are obviously tracking the patrol.
- The unusual absence of pedestrian traffic and people on porches.
- Stores, markets or street vendors closed suddenly or without explanation.
- Changes in civilian attitude toward patrol members.
- Unknown individuals or vehicles in the patrol area.
- Unfamiliar vehicles parked in the patrol area (possible car bomb).
- Roadblocks.
- Children throwing rocks at patrols to possibly draw the patrol's attention away from a more serious danger, such as a deliberate ambush.
- Vehicles riding unusually low due to overloading (possibly ferrying people, weapons, explosives).
- Agitators trying to provoke an incident with patrol members.

- Absence of the usual stray dogs (dogs are adept at sensing danger and avoiding it).
- Anti-American graffiti suddenly appearing in the patrol area.
- Pictures of enemy leaders and martyrs posted in the patrol area.
- Civilian workers failing to appear at U.S. or friendly bases.
- Normal deliveries and pick-ups conducted late or early without reason.
- Sudden change of civilian sentiment in newspaper articles, radio broadcasts or other media.
- Women and children leaving to live elsewhere.

Immediate Actions Upon Enemy Contact

Reaction to Sniping

Snipings are often executed from a single firing point, but coordinated snipings delievered from multiple points are not uncommon.

Patrol element or team leaders should constantly try to identify likely firing points and anticipate their own reactions to a possible shooting. Normally, snipers in an urban environment have a detailed withdrawal plan. Once a patrol comes into contact with a sniper, the patrol leader must immediately assess the situation and maneuver his patrol accordingly. The patrol's mission, location, size, ROE, and location of the threat often determine whether the patrol will attempt to neutralize the targeted sniper. If the patrol leader decides to kill or capture the sniper, he uses planned and rehearsed immediate actions to maneuver and counter the sniper's assault. The goal is to kill the sniper or cut off his escape and capture him. There are three immediate reactions to neutralizing a sniper: initial contact, immediate follow-up, and subsequent follow-up.

Initial Contact. The initial contact is made when the sniper fires the first shot. The patrol must react immediately and positively to get behind the firing position in order to kill or capture the gunman. The period of contact ends when the gunman is killed or captured, or the patrol element or team leader on the scene ends it. The following technique is the same for both squad- and platoon-sized patrols:

- The element or team in contact attempts to identify the firing position and maneuvers designated marksman into position to return well-aimed and controlled fire. Other members of the patrol take up positions to cover the marksman's engagement.
- The patrol element or team leader in contact sends initial contact report to the patrol leader, who notifies higher headquarters.
- The element or team leader in contact determines appropriate cut-off positions and relays them to flanking elements or teams.
- The patrol element or team leader in contact continues to observe the firing point, but does not enter it due to the possibility of booby traps. Flank elements or teams set up along likely escape routes.
- The incident ends when either the sniper ceases fire or is neutralized.

Immediate Follow-Up. Regardless of the fate of the gunman, isolation of the firing point is necessary to prevent reinforcement and preserve forensic evidence (scent, spent casings, etc.). If not under fire, members of the patrol element or team cordon off the area surrounding the firing point. Flanking elements or teams maintain their positions and prevent civilians from entering the area. The patrol leader moves to link-up with the element or team in contact (if not his own), and makes an estimate of the situation. The patrol leader sends a SPOTREP to higher headquarters.

Subsequent Follow-Up. The aim of the subsequent follow-up is to use follow-on forces to clear the building of remaining resistance or to obtain evidence that can be used to capture the gunman. The patrol leader establishes a position where he can brief arriving units (reaction force commander, S-2 representative, EOD personnel, etc.). Once the arriving units have been briefed, recommendations are made to higher headquarters via radio. No one is allowed into the cordon without the patrol leader's approval.

Reaction to Becoming Decisively Engaged

If a patrol becomes decisively engaged from numerous firing positions, the following immediate action should be taken:

- All patrol members move to available cover and return accurate fire on identified firing points.
- The patrol leader assesses the situation and makes a decision to either request the reaction force or break contact.
- If the reaction force is requested, the patrol will maintain its position until the reaction force arrives. The patrol should use fire and maneuver to gain better tactical positioning and support the arrival of the reaction force. When the reaction force arrives,

its commander may decide to either clear occupied buildings or cover the patrol during its extract.

Reaction to Bomb Threat or Discovery

The use of command-detonated explosive devices is a common ambush tactic employed by a terrorist or insurgent in an urban environment. The appropriate response to a reported threat or an actual discovery is generally involves four steps (known as the four Cs)—

1. The patrol leader **CONFIRMS** the presence of the suspicious item.

2. Without touching or moving anything suspicious, patrol elements or teams **CLEAR** the immediate danger area to a minimum of 100 meters. The area is cleared from the suspected device outward, inform civilians as to the reason for evacuation.

3. A&S teams establish a **CORDON** to secure the cleared area. Avenues of approach are cordoned off to keep people out and to protect EOD or engineer personnel clearing the device. The assistant patrol leader acts as the cordon commander and informs the patrol leader when the cordon is secure. An effective cordon technique is to tape off the area with engineer tape, creating both a physical and psychological boundary.

4. **CONTROL** of the area is maintained throughout the bomb clearing operation by the patrol leader. The patrol leader sends a report to a superior concerning details of the device (if known) and the area affected. The patrol leader coordinates with arriving personnel (EOD, engineers, etc.). The patrol leader maintains communication with the assistant patrol leader and keeps the Marines informed of the progress of the clearing operation.

Reaction to a Bomb Detonation

Bombs may be used by an insurgent as a means of initiating an ambush on mounted or dismounted patrols, in which case the actions for decisive engagement apply. Immediate action in response to an isolated explosion is similar to that used in reaction to a sniping and breaks down into the same three phases:

Initial Contact. The patrol leader attempts to identify the likely initiation point and sends an initial contact report to higher headquarters. If the bomb was command-detonated, the patrol leader sends his A&S teams deep to cut off the bombers' escape routes. Any casualties are moved a minimum of 100 meters from explosion and out of the line of sight to it.

Immediate Follow-Up. The A&S teams may need to maneuver to positions behind the likely initiation point to cut off escape. Once in position, personnel checks are conducted and any suspects are detained. The patrol leader coordinates requests for required support (MEDEVAC, reaction force, etc.)

Subsequent Follow-Up. Due to the possibility of secondary detonations, the four Cs (confirm, clear, cordon, and control) can be conducted as in reaction to a bomb discovery or bomb threat.

Civil Disturbances

Urban patrols must prepare to react to spontaneous aggression by the local populace. In many cases, civil disturbances are organized by the enemy to draw dismounted patrols into a targeted area, or to distract them from enemy activity occurring elsewhere. Civil disturbances are generally divided into two categories: minor aggressive actions, and full-scale rioting.

Minor aggressive actions are activities characterized by rock-throwing or use of devices such as Molotov cocktails and may either be directed at the patrol or take place between different ethnic factions of the population. Minor aggressive actions are normally spontaneous in nature and may have minimal or limited objectives for the insurgents.

Full-scale rioting events are usually in response to another major event or incident that may enflame the populace. Full-scale riots are well-planned and orchestrated, with clear objectives or targets in mind. At times, patrols will need to attempt to maintain control of a civil disturbance situation; however, dismounted and mobile small unit patrols should generally avoid potential flashpoints. Procedures to handle civil disturbances are as follows:

- The patrol leader reports the incident to headquarters and attempts to diffuse the crowd by talking to crowd leaders.
- If the patrol leader determines the size of the disturbance is too large for the force to handle, the patrol should move away from the disturbance to a safer, more remote covered area and occupy positions to observe and report the situation to higher headquarters. To prevent the patrol from being pursued by the crowd, the patrol should move quickly and change direction, often at road junctions, to gain distance from it.

Patrol members should maintain dispersion to create a more difficult target. They should face the crowd at all times to see and avoid any projectiles thrown. Individual self-discipline must be maintained throughout the disturbance. Marines charging into the crowd or throwing objects back at the crowd will only worsen the situation. If pursued or trapped, the patrol leader may consider using riot control agents (combat support, pepper gas, etc.) to disperse or slow a crowd's movement. The use of riot control agents must be authorized under the established ROE. If the patrol leader believes the situation is deteriorating beyond the patrol's ability to control it, the patrol leader should request the reaction force, which may be better equipped to handle a large riot or mob.

Break Contact

As with patrols in rural areas, the patrol leader may be forced to break contact as a result of decisive engagement with the enemy. On the basis of his estimate of the situation, the patrol leader will normally break contact in one of the following ways:

- As a patrol, with elements providing cover for movement as defined by clock direction and distance.
- As individual units/teams taking separate routes out of the area, then linking up at a designated rally point a safe distance away from the engagement.

As in any contact with enemy forces, smoke may be employed to screen movement. Fire support agencies can be utilized to suppress targets; riot control agents can be employed to disrupt enemy movement.

Reentry of Friendly Lines

The reentry of a dismounted patrol into an urban patrol base is no different from that of a patrol conducted in a rural area. The same planning considerations and control methods apply.

Missions Related to Urban Patrolling

House Calls

House calls missions are usually part of a coordinated effort to collect information within the area of operations. They involve obtaining up-to-date information on particular houses and occupants. When possible, local police should accompany patrols to do the actual talking to the occupants. If this is not possible, a technique that may be used by an urban patrol is—

- A&S elements move to provide cover around the target house.
- The headquarters element provides security just outside the house.
- The patrol leader and one Marine for security enter the house, if invited. If not invited, they talk to the occupants on the doorstep. Patience and tact are required in requesting information. An interpreter should be present when language differences exist.

Vehicle Checkpoints

Urban terrorists or insurgents commonly use vehicles to transport personnel, weapons, explosives, and equipment. Civilian vehicles are often used for these purposes, creating the requirement to check as many vehicles as possible. While permanent, fortified checkpoints may be conducted along approaches into an urbanized area, dismounted patrols can be employed to establish hasty vehicle checkpoints to stop vehicles and to keep the enemy off-guard. The two common types of vehicle checkpoints patrols establish are hasty and deliberate.

Hasty checkpoints are deployed anywhere based upon the decision of the patrol leader. Patrols must not set patterns through the frequent use of the same sites.

Deliberate checkpoints are tasked by higher headquarters to achieve a specific purpose. Time and locations are carefully considered to avoid setting patterns.

The general layout for a squad-sized, two-way dismounted checkpoint is depicted in figure 13-8 on page 13-20. The technique is as follows:

- The A&S teams are positioned stealthily in blocking positions on both sides of the road.
- Both the patrol leader and assistant patrol leader act as "talkers" for each direction of traffic (with local police or interrogator or translator Marines acting as interpreters) while a two-man team from the headquarters element physically searches the vehicles.
- Obstacles or parked vehicles may be employed to create a staggered roadblock in center of the checkpoint to slow approaching vehicles.
- The checkpoint location should be sited so that approaching vehicles cannot see it until they have passed a security team, and they have no escape route then available.
- Signs announcing the checkpoint should be displayed a safe distance from the search area for safety to both drivers and Marines.
- Normally higher headquarters will issue criteria that determines which vehicles are searched, but random checks of cars should normally be made as

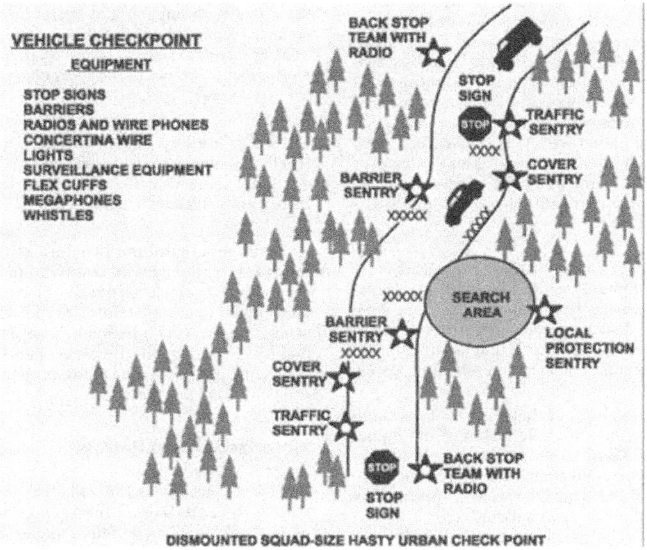

Figure 13-8. Dismounted Squad-sized Hasty Urban Vehicle Checkpoint.

well. While the vehicle is being searched the driver should accompany the searcher around the vehicle.

- Vehicle occupants should be made to exit the vehicle and then searched. Whenever possible, women should be searched by female police or female Marines.
- All patrol members should conduct themselves with courtesy and politeness. If nothing is found, an apology for the inconvenience is recommended.
- A hasty vehicle checkpoint should not be conducted any longer than 30 minutes for security reasons.
- The ROE should dictate whether or not action should be taken against vehicles that fail to stop at the checkpoint. Failure of a vehicle to stop does not automatically give authorization to fire.

Observation Posts

Urban observation posts are established to provide extended security, not only for patrol bases but also for patrols operating within the observation post's sector of observation. Observation posts can be established in conjunction with sniper operations and for providing overwatch for patrols operating within their sector of observation. Observation posts are nor-mally positioned on dominating terrain or in buildings outside the patrol base itself.

Insertion to the observation posts and conduct of observation may be either overt or clandestine in nature. Overt observation posts usually will be hardened positions to increase security. A patrol provides cover while the observation post is being inserted. A clandestine observation post relies on stealth of insertion and occupation for protection. It is normally positioned in abandoned buildings to cover sectors of observation that overt observation posts cannot. Because of their nature, clandestine observation posts are difficult to successfully establish and should not be manned for an extended period of time. Orders establishing observation posts (and patrols) must address the method of extraction as well as actions upon compromise/attack.

Cordon and Search

The cordon and search mission involves isolating a predesignated area by cordoning it off and systematically searching for enemy personnel, weapons, supplies, explosives or communications equipment. While large-scale cordon and search operations are planned and rehearsed in advance and normally entail extensive coordination with local law enforcement agencies, a squad-sized urban patrol may often conduct a cordon and search of a point target—searching one house or building identified by intelligence as a possible weapons cache.

The basic principle of a search of a populated area is to conduct it with limited inconvenience to the population. The populace may be inconvenienced to the point where they will discourage urban guerillas or insurgents from remaining in the area, but not to the point that they will assist the enemy as a result of the search.

Upon receiving intelligence that warrants the searching of a building or a specific tasking from higher headquarters, A&S elements of the patrol move to establish an inner cordon around the target building to seal it off, with the primary intent of preventing movement out of the targeted building.

On order, the designated reaction force deploys to establish an outer cordon, oriented outward some distance from the inner cordon and covering routes leading into the area in order to prevent outside interference/reinforcement. The reaction force maintains a reserve to reinforce either cordon or react to unfolding events (civil disturbance in response to the operation).

Once the cordons have been established, the patrol leader, with the assistance of local police or interpreters, informs the local populace that a building is about to be searched, that a house curfew is in effect (if permitted by higher headquarters), and that all occupants should remain indoors. Occupants of the target house are instructed to gather at a central location to stay out of the way of the search party.

The headquarters element, having linked up with any required assistance (explosive ordnance disposal (EOD), ITT, etc.) now acts as the search party and accompanies local police. A female searcher should be included in the party, if necessary.

Occupants are searched and screened first for possible enemy personnel. Apprehended persons are evacuated as soon as possible.

The head of the household should accompany the search party throughout the operation to be able to counter incriminating evidence and possible accusations of theft and looting against Marines. If possible, a prominent member of the local community should act as a witness.

Buildings are best searched from top to bottom. Ideally, the search is conducted with the assistance of combat engineers using mine detectors to locate hidden arms and ammunition.

If the targeted building is empty or the occupant refuses entry, it may be necessary to forcefully enter the premises to conduct a search of the dwelling. If an unoccupied house containing property is searched, arrangements should be made with the local community to secure it until its occupants return. Unnecessary force and damage to property should be avoided during the search.

Motorized Urban Patrols

The advantages of a motorized urban patrols is their ability to capitalize on the speed, mobility, and protection offered by various vehicles. They may be motorized, mechanized or armored vehicles or a combination. Generally, motorized urban patrols possess greater combat power than dismounted patrols and can cover larger areas faster than dismounted patrols.

The disadvantages of motorized urban patrols is that they are restricted to roads and are vulnerable to ambush by the enemy. They are also restricted in their ability to interact with the local populace.

Motorized patrols are generally organized in the same manner as dismounted patrols (see fig. 13-9). Unit integrity is maintained when assigning personnel to specific vehicles.

The urban patrolling principles apply to motorized patrols in much the same manner as dismounted patrols. Mutual support and depth are achieved by maintaining constant observation between vehicles and coordinating support with any dismounted patrols in the area. All-around security is achieved through the use of constant observation as well as the vehicle's mobility and firepower. Positive communications between units or teams are maintained through vehicle radios. Patrol routes and speeds are varied to promote deception or pattern avoidance.

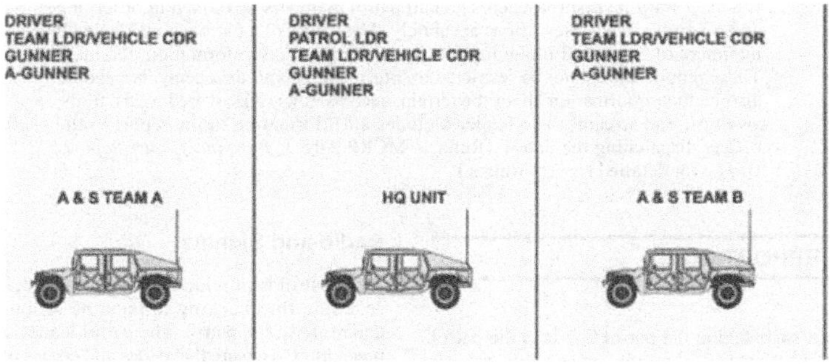

Figure 13-9. Organization of a Squad-sized Motorized Patrol.

The canalizing nature of streets and alleys force vehicular patrols to use a traveling overwatch movement technique to reduce vulnerability to ambushes. All vehicles travel at a moderate rate of speed with the lead vehicle stopping only to investigate potential danger areas. If vehicles must stop in danger areas, designated crew members' will dismount to provide security. The gunner will remain at the ready and in the turret while the driver remains in the driver's seat with the vehicle running.

Vehicles should move at a high rate of speed only when responding to an incident. At all other times, vehicle speed should be between 15 to 20 mph to allow for quick reaction and good observation. Distances between vehicles should be approximately 50 meters (one half to one city block) or such that visual contact and mutual support are ensured. Particular care is taken at major road junctions and other danger areas to ensure individual vehicles do not become isolated.

Vehicles with doors removed generally enhance observation and overall security, yet expose Marines to thrown objects, theft and concealment.

CHAPTER 14. INFORMATION AND REPORTS

It is necessary that patrol leaders and all patrol members be trained in observing and reporting their observations accurately. The leader of a patrol should have all members of the patrol immediately signal or report any information obtained. These reports should not be restricted to information about the enemy, but should also include information about the terrain, such as newly discovered roads, trails, swamps, and streams. The leader includes all information in the report to the officer dispatching the patrol. (Refer to MCRP 2-15.3, *Reconnaissance Reports Guide,* for detailed report formats.)

14001. REPORTING

The officer dispatching the patrol instructs the patrol leader on whether and when messages are to be sent back during the patrol and what communication means to use. Messages may be oral or written. They must be accurate, clear, and complete. Every message should answer the question what, where, and when. For a detailed discussion on reporting, see paragraph 6002.

Verbal Messages

A patrol leader sending a verbal message should make it simple, brief, and avoid using numbers and names. The messenger should accurately repeat the message back to the patrol leader before leaving.

Written Messages

In preparing written messages, the patrol leader must distinguish between fact and opinion. Information about the enemy should include: strength; armament and equipment; actions; location and direction of movement; unit destination, if known; time enemy was observed; and the patrol's location when the observation was made. Use of an overlay or sketch may often simplify the message.

Messengers

A messenger team is given exact instructions as to where to deliver the message and the route to take. Any information obtained along the route should be reported at the time the message is delivered. Messengers must be given all practical assistance. If in danger of capture, the messenger immediately destroys the message.

Radio and Signals

If the patrol is provided with a radio, a definite radio schedule for checking in must be arranged before departure of the patrol. The patrol leader takes every precaution to ensure that codes and copies of messages are not captured by the enemy. If a close reconnaissance of enemy lines is required, the radio should be left in a concealed location at a safe distance from the enemy. Once a report is sent by radio, the patrol should immediately leave the area to avoid the possibility of detection by enemy locating devices. Pyrotechnics (flares, colored smoke, grenades) and air-ground panels may also be used by patrols for reporting information by a prearranged signal.

SALUTE Report

Information must be reported as quickly, accurately, and as completely as possible. An established method to remember how and what to report about the enemy is to use the acronym SALUTE:

Size
Activity
Location
Unit
Time
Equipment

An example of such a report is: "Seven enemy soldiers, unit unknown, traveling SW, crossed road junction on BLACK RIDGE at 211300 August carrying one machine gun and one rocket launcher."

14002. CAPTURED ITEMS

Every patrol should make a practice of searching enemy casualties, prisoners, and installations first for booby traps, then for equipment, papers, maps, messages, orders, diaries, and codes. Search techniques should be practiced often in order to reduce the time exposed to potentially dangerous situations. Items found are collected by the patrol leader and turned in with the patrol report. The items found are marked as to time and place of capture. When possible, captured items should be linked to a specific prisoner who possessed the items or to the place where the items were found. When this is done, the enemy prisoner of war tag and item tag are marked accordingly. The patrol leader must impress upon the members of the patrol the importance of turning in all documents and equipment. Furthermore, the patrol leader must ensure all information gathered by the patrol is quickly disseminated.

14003. PRISONERS

A patrol normally does not capture prisoners unless required by the mission. If prisoners are taken, the "5S and T" rule applies. This memory aid stands for: search, segregate, silence, speed, safeguard, and tag.

Search. Prisoners are body-searched thoroughly for weapons and documents as soon as they have been captured. This search must include the helmet, body armor, and gas mask. These items are left with the prisoner for protection until the patrol is completed. Weapons, equipment, and documents are tagged and immediately sent to the patrol leader.

Segregate. Prisoners are segregated into isolated groups: officers, noncommissioned officers (NCOs), privates, deserters, and civilians. By segregating prisoners, it makes it more difficult for leaders to organize escapes and issue orders to subordinates.

Silence. Silence is essential. Do not allow prisoners to talk to each other.

Speed. Speed is required in getting prisoners to the commander who dispatched the patrol. Timely information secured from prisoners is essential.

Safeguard. Prisoners are safeguarded as they are moved. They are restrained, but not abused. If the patrol will soon reach friendly positions, prisoners are not given cigarettes, food or water until they have been questioned by interrogators. If the patrol will not return to a friendly position for a lengthy time, the prisoners are given food, water and medical aid.

Tag. Equipment and personal effects recovered from prisoners are tagged so that they may later be matched back to specific individuals.

14004. PATROL REPORT

Every patrol leader makes a report when the patrol returns. Unless otherwise directed, the report is made to the person ordering the patrol. If the situation permits, the report is written and supported by overlays and/or sketches. The patrol leader's report should be a complete account of everything of military importance observed or encountered by the patrol while on the assigned mission. It should include the following information:

- Size and composition of patrol.
- Tasks and purpose (mission).
- Time of departure.
- Time of return.
- Routes, out and back (show by sketch, azimuth, trace on map).
- Terrain (general description to include any man-made or natural obstacles and critical terrain features which, if occupied by either enemy or friendly forces, would allow them to control the surrounding area).
- Enemy (size, activity, location, unit, time, equipment).
- Any map corrections (show on map).
- Miscellaneous information not covered elsewhere in report.
- Results of enemy encounters.
- Condition of patrol, including disposition of any dead or wounded.
- Conclusion and recommendations.

14005. PATROL CRITIQUE

After the patrol has rested and eaten, the patrol leader should hold a critique. Constructive criticism is made. It is an excellent time to prepare for future patrols by going over lessons learned as a result of the patrol.

Appendix A. Patrol Warning Order

The warning order is issued as soon as practical with all available information included to assist patrol members in preparation.

1. Situation. Friendly and enemy situation information necessary for initial preparation.

2. Mission. Statement of what the patrol is to accomplish, and the purpose for accomplishing it. When, how, and where will be discussed in the patrol order.

3. Execution

a. Task Organization. General patrol organization and assignment of responsibilities if known; otherwise, promulgate in the patrol order.

b. Tasks. Alert subordinate leaders to patrol tasks (e.g., stream crossing, helicopter rappelling, demolitions) requiring preparation prior to departure.

c. Coordinating Instructions

(1) Time schedule and location for individual preparation, rest, briefings, in-spection, rehearsal, and departure.

(2) Time, place, uniform, and equipment for receiving the patrol order.

(3) Tasks for subordinate leaders to direct and supervise the initial preparation which may include drawing ammunition, rations, and special equipment; conducting immediate action drills or other necessary individual or unit training; meeting and briefing attachment personnel; reconnoitering the area for passage of lines; and coordinating with the necessary unit leaders.

(4) Preliminary guidance to specialists and key individuals regarding their roles and organization within the patrol.

4. Administration and Logistics

a. Individual uniform, equipment, weapons, and prescribed load of rations, water, and ammunition.

b. Crew-served weapons (if required) and guidance regarding distribution of weapons and ammunition during movement.

c. Special equipment requirements (wire cutters, demolitions, radios, flashlights, infrared equipment, mines, binoculars) and their distribution during movement.

d. Restricted or prohibited items.

5. Command and Signal

a. Designation of assistant patrol leader and his role in preparation.

b. Designation of navigators and radio operators (if required).

c. Brief outline of patrol leader's schedule for preparation, and where he can be reached.

APPENDIX B. PATROL ORDER

A patrol order follows a warning order. Any of the following subjects that have been addressed in the warning order may be omitted with the exception of the mission statement. A patrol order is more detailed than a 5-paragraph squad order in that a great deal of attention is given to individual duties.

The patrol order also provides orientation information, both enemy and friendly, that impacts the patrol and enemy forces. This information can include, but is not limited to, weather, terrain, visibility, NBC considerations, local population situation, terrain model and/or map orientation, and behavior.

1. <u>Situation</u>

   a. <u>Enemy Forces</u>

      (1) Composition, disposition, and strength are based on size, activity, location, unit, time, and equipment (SALUTE).

      (2) Capabilities and limitations to defend, reinforce, attack, withdraw, and delay (DRAW-D).

      (3) Enemy's most probable course of action.

   b. <u>Friendly Forces</u>

      (1) Mission of next higher unit (task and commander's intent).

      (2) Adjacent unit missions (task and intent). Also identify left, front, right, and rear.

   c. <u>Attachments and Detachments</u>. (date and time effective).

2. <u>Mission</u>. The mission is the task to be accomplished, and its purpose (who, what, where, when, and why). For patrols, specify if the mission or time has priority.

3. <u>Execution</u>

   a. <u>Commander's Intent and Concept of Operations</u>

      (1) Commander's intent.

      (2) The concept of operations tells the where, how, and who and lays out the patrol leader's general scheme of maneuver and fire support plan. It outlines the following:

(a) Task organization of the patrol.

(b) Movement to the objective area, to include navigation method.

(c) Actions in the objective area.

(d) The return movement, to include navigation method.

(e) Use of supporting forces (including illumination, if required).

b. Subordinate Element Missions. Subordinate element missions (task and purpose) are assigned to elements, teams, and individuals, as required.

c. Coordinating Instructions. This paragraph contains instructions common to two or more elements, coordinating details, and control measures applicable to the patrol as a whole. At a minimum, it includes—

(1) Time of assembly in the assembly area.

(2) Time of inspections and rehearsals (if not already conducted).

(3) Time of departure and estimated time of return.

(4) Location of departure and reentry of friendly lines and the actions associated with departure and reentry.

(5) Details on the primary and alternate routes to and from the objective area.

(6) Details on formations and order of movement.

(7) Rally points and actions at rally points.

(8) Final preparation position and actions at this position.

(9) Objective rally point and actions at this point.

(10) Actions at danger areas.

(11) Actions in the event of enemy contact.

(12) Details on actions in the objective area not covered elsewhere.

(13) Estimated time of patrol debriefing upon return.

4. Administration and Logistics

a. Changes/additions to uniform, equipment, and prescribed loads from that given in the warning order.

b. Instructions for handling wounded and prisoners.

5. Command and Signal

a. Command Relationships. Identify key leaders and chain of command.

b. Signal. Challenge and password, arm and hand signals, special signals, and radio frequencies and call signs.

APPENDIX C. PATROL EVALUATION CHECKLIST

This appendix provides a comprehensive checklist of critical patrolling steps, techniques, and procedures to aid unit leaders to critique patrol performance during training. Unit leaders may use the list as they observe the performance of a patrol to provide the patrol leader with a detailed analysis of the performance.

Warning Order	Yes	No
Did the PL ensure all patrol members were present before issuing the warning order?		
Did the PL issue a brief statement on the enemy situation?		
Did the PL issue a brief statement on the friendly situation?		
Did the PL state the mission in a clear, positive manner?		
Did the PL list all members of the patrol including attachments?		
Was the chain of command for the patrol covered fully?		
Were all members of the patrol asigned positions and duties in particular squads and teams?		
Were all the necessary individual duties assigned?		
Did the PL follow established principles in organizing the patrol into squads and teams?		
Was each patrol member assigned a particular weapon to carry on patrol?		
Did the PL list all special equipment needed to accomplish the mission?		
Was required special equipment assigned to the proper element to carry?		
Did the PL select uniform and equiment common to all based on METT-T?		
Coordination with Adjacent Units		
Did the PL coordinate with other patrols operating to the right and left?		
Was the route out and back coordinated?		
Was the time of departure and return cocordinated?		
Were call signs and frequencies coordinated?		
Was a signal for the FPF coordinated so as not to approach friendly lines during this time?		
Coordination With Front Line Units		
Did the coordinator pass the size of the patrol?		
Was the time of departure and return coordinated?		
Did the coordinator give a general area of oeprations for the patrol?		
Did the coordinator ask for information on known or suspected enemy positions and/or obstacles?		
Did the coordinator ask about information on the latest enemy activity?		
Did the coordinator ask for detailed information on friendly fire support available and the unit's barrier plan?		
Was the location of the IRP established and coordinated?		
Did the coordinator ask the forward unit to monitor their patrol frequency?		
Was the current challenge and password confirmed?		
Did the coordinator request that all information coordinated be passed on to any relieving unti?		
Patrol Order		
General		
Did the PL check to ensure all patrol members were present before issuing the patrol order?		

	Yes	No
Did the PL issue the patrol order in a forceful, confident manner?		
Did the PL make maximum use of available visual aids in issuing the patrol order (i.e., terrain model, sand table, map board, chalkboard)?		
Did the PL issue the patrol order in correct sequence?		
Did the PL issue the entire patrol order without allowing interruptions by patrol members?		
Did the PL adequately answer all questions asked by patrol members		
Did paragraph 1a (Enemy Situation) include—		
A weather forecast for the period of operation?		
A description of the terrain over which the patrol was to operate?		
Identification or description of enemy units known to bein the area of operations?		
Known locations of enemy units?		
Recent activity of enemy units?		
Stength of enemy units in the operating area?		
Did paragraph 1b (Friendly Situation) include—		
Mission of the next higher unit?		
Locations, missions, and planned actions of units on right and left?		
Fire support available to support the patrol (as per the coordination)?		
Missions and routes of other patrols operating in the immediate area?		
Did paragraph 1c (Commander's Intent) include—		
All views and ides of what the commander wants to be accomplished?		
The final result desired by the commander?		
Did paragraph 1c (Attachments and Detachments) include—		
All attachments to patrol and effective time of attachment?		
All detachments from patrol and effective time of detachment?		
Did paragraph 2 (Mission) include, at a minimum—		
Who was to conduct the patrol?		
What the patrol was tasked to do (i.e., conduct a point recon patrol)?		
Where the action was to take place?		
Why the action was to take place (i.e., the purpose)?		
Did paragraph 3 a (Commander's Intent and Concept of Operations) include—		
A complete concept of operations?		
A detailed description of the mission of elements?		
A detailed description of the mission of teams?		
A detailed description of the duties of specific individuals (i.e., navigator, compass man, pace m an, corpsman, APL)?		
Time of departure and time of return?		

	Yes	No
Type of formations and order of movement to be used?		
Description of the route and alternate routes(s) to include azimuths and distances between checkpoints as a minimum?		
Techniques to be used in the departure from friendly areas?		
Techniques to be used for the reentry into friendly areas?		
Location of the IRP and a tentative ORP, to include grid coordinates and recognizable terrain features?		
Method for designating and passing rally points?		
Actions to be taken at rally points if their use became necessary?		
Actions to be taken in the event of enemy contact?		
Actions to be taken at danger areas?		
Complete, detailed description of actions to be taken at the objective?		
Times and locations for rehearsals and the order of priority for rehearsals?		
Times and Icoations for inspections and the methods of conduct (to include uniform and equipment to be worn and/or carried)?		
Where, when, and by whom the debriefing is to be conducted?		
Did paragraph 4 (Administration and Logistics) include—		
Rations to be carried or reference to the warning order if there is no change?		
Arms and ammunition to be carried or reference to the warning order if there is no change?		
Uniform and equipment to be worn and carried or reference to the warning order if there is no change?		
Method for handling wounded or dead?		
Method for handling prisoners?		
Did paragraph 5a (Signal) include—		
Complete description of all signals, code words, frequencies, and call signs to be used within the patrol?		
Were signals adequate?		
Radio call signs to be used?		
Primary and alternate frequencies to be used?		
Call signs and frequencies to be used with other units in the area of oeprations (if applicable)?		
Required reports to be transmitted to higher headquarters, along with occasions for reporting?		
Code words and brevity codes to be used between the patrol and higher headquarters?		
Current challenge and password to be used in friendly held areas?		
Current challenge and password to be used forward of FEBA?		
Did paragraph 5b (Command) include—		
Chain of command or reference to the warning order if there is not change?		
Location of the PL during all stages of operation?		

	Yes	No
Location of the APL during all stages of operation?		
Inspection Rehearsal		
General		
Did the P L adequately use the allotted inspection time?		
Was the inspection conducted in an orderly manner?		
During the inspection, did the PL check the following:		
Completeness and correctness of uniform?		
Completeness of all equipment necessary to accomplish the mission?		
Operational condition of equipment?		
Did the PL question members to ensure they knew—		
The mission of the patrol?		
The concept of operations?		
Their individual duties and responsibilities?		
Chain of command and succession?		
Duties and responsibilities of key personnel?		
Did the PL utilize the rehearsal area to conform as much as possible to area of operations?		
Were the following major actions rehearsed:		
Actions at the objective?		
Actions at danger areas?		
Actions on enemy contacts?		
Departure and reentry of friendly lines?		
Did the PL critique each action after it had been rehearsed?		
Were interpatrol communications and control measures checked at rehearsal?		
Did the PL maintain control of the patrol during the rehearsal?		
Movement		
Did the PL use proper formations for movement?		
Did the PL adequately control the point team?		
Did the PL check the compass men?		
Was the PL aware of his position at all times?		
Was light and noise discipline enforced?		
Was the rate of movement appropriate for the mission?		
Did the PL make full use of pace?		
Did the PL make full use of count?		
Was proper security maintained during movement?		
Did the PL select and pass proper rallying points?		
Did the PL ensure that all members knew location of rallying points?		

	Yes	No
Did the PL properly use arm and hand signals?		
Did the PL recognize and halt the patrol a safe distance from a danger area?		
If the PL conducted a map check, did he conduct it properly (light discipline, security, etc.)		
Did designated individuals reconnoiter the far side of a danger area properly?		
Were support teams emplaced properly?		
Did the PL use a proper formation in crossing the danger area?		
Were reports made on enemy contact and at checkpoints?		
Actions on Enemy Contact		
Did the PL halt patrol a safe distance from a tentative ORP?		
Did the PL issue an adequate frag order before departing to look for a tentative ORP?		
Did the PL take appropriate personnel to look for the ORP?		
Did the PL select a suitable ORP?		
Did the PL secure the site adequately?		
Did the patrol occupy the ORP as stated in the patrol order or subsequent frag order?		
Did the patrol occupy the ORP in an orderly manner?		
Did the PL maintain control of the patrol during occupation of the ORP?		
Did the PL issue a satisfactory contingency plan to the APL before departing on a leader's recon?		
Did the PL take appropriate personnel on a leader's recon?		
Did the PL maintain or ensure that the patrol leader had communications with the patrol?		
Did the patrol avoid being detected by the enemy during the leader's recon?		
If contact was made, did the PL take appropriate action?		
Was security maintained during the leader's recon?		
Did the PL ensure the objective was kept under surveillance?		
Were appropriate orders given to surveillance teams?		
If a leader recon proved a tentative ORP unsuitable, did the PL move the ORP?		
Was a satisfactory leader's recon conducted?		
Was the ORP move conducted in a satisfactory and orderly manner?		
On return to the ORP, did the PL issue appropriate frag order and allow enough time for dissemination?		
Actions at Objective		
Did the PL issue a frag order for action at the objective?		
Was security emplaced prior to the teams departing from the ORP?		
Did the PL employ the terrain at the objective to the best advantage?		
Was stealth maintained while moving into the objective?		

	Yes	No
Did the Pl exercise positive control of squads, teams, and individuals at the objective?		
Were actions at the objective in accordance with the details outlined in the patrol order?		
Did the PL make use of supporting arms at the objective?		
Was the action at the objective successful?		
Was withdrawal from the objective accomplished quickly and orderly?		
Did units withdraw according to the details reflected in the patrol order?		
Was reorganization at the ORP completed in an expeditious and orderly manner?		
Did the patrol withdraw from the ORP quickly and quietly?		

Appendix D. Ambush Formations

Formations discussed here are identified by names which correspond to the general pattern formed on the ground by deployment of the assault element.

Line Formation

The assault element is deployed generally parallel to the route of movement of the enemy. This positions the assault element parallel to the long axis of the killing zone and subjects the target to heavy flanking fire. The size of the force which can be trapped in the killing zone is limited by the area the assault element can effectively cover. The enemy is trapped in the killing zone by natural obstacles, mines, demolitions, and direct fires. (See fig. D-l.)

A disadvantage of the line formation is the chance that lateral dispersion of the target may be too great for effective coverage. The line formation is appropriate in close terrain that restricts enemy maneuver, and in open terrain where one flank is protected by natural obstacles or can be protected by mines and demolitions. Similar obstacles can be placed between the attack force and the killing zone to provide protection from possible enemy counter attack. When an ambush is deployed in this manner, access lanes are left so that the enemy can be assaulted. (See fig. D-2.)

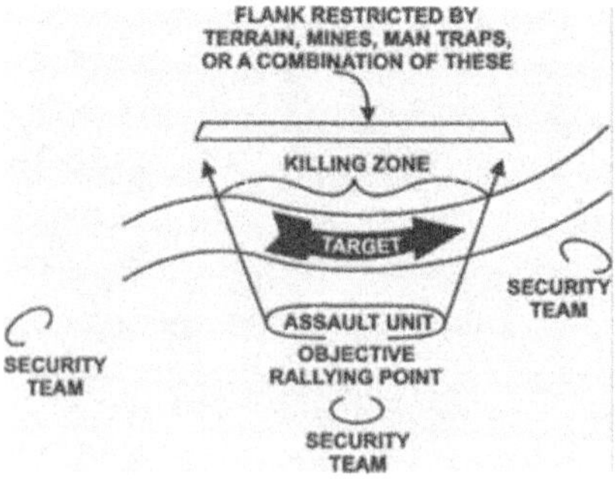

Figure D-1. Line Formation (Harassing or Destruction Ambush).

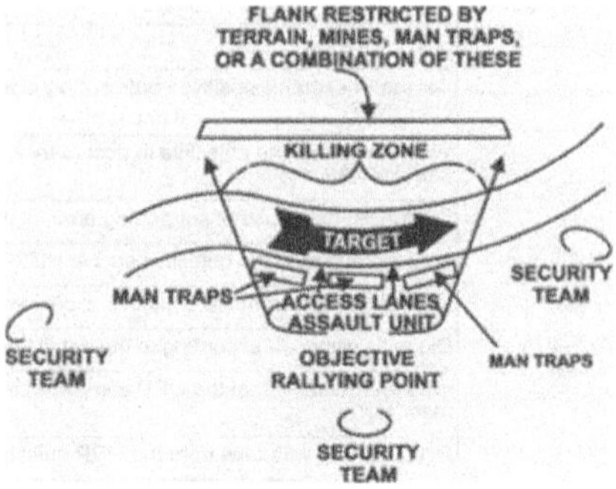

Figure D-2. Line Formation (Access Lanes for Assault of Target).

The main advantage of the line formation is its relative ease of control under all conditions of visibility.

L Formation

The "L" shaped formation is a variation of the line formation. This formation is very flexible because it can be established on a straight stretch of a trail or road (see fig. D-3) or at a sharp bend in a trail or a road (See fig. D-4 on page D-2).The long side of the assault element is parallel to the killing zone and delivers flanking fire. The short side of the attack force is at the

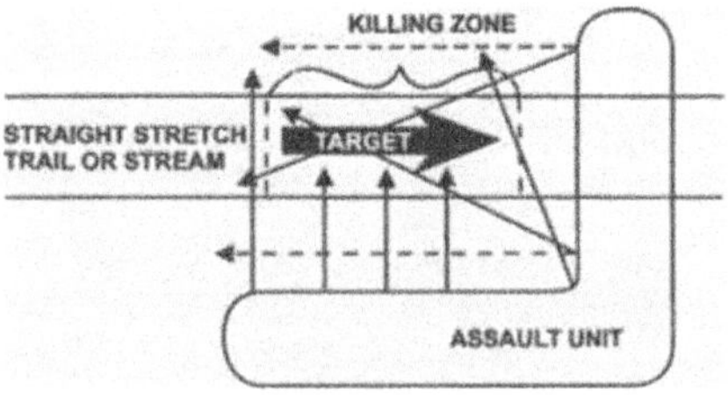

Figure D-3. "L" Formation (Destruction Ambush).

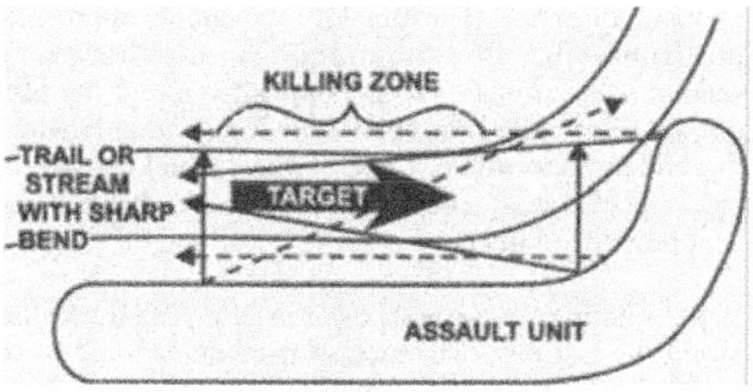

Figure D-4. "L" Formation (Bend of Trail or Stream).

end of, and at right angles to, the killing zone and delivers enfilading fire that interlocks with fire from the long side. When appropriate, fire from the short side can be shifted to parallel the long side if the enemy attempts to assault or escape in the opposite direction. In addition, the short side prevents escape and reinforcement. (See fig. D-5.)

"Z" Formation

The "Z" shaped formation is another variation of the line formation.

The assault element is deployed as in the "L" formation, but with an additional side so that the formation resembles a "Z". (See fig. D-6).

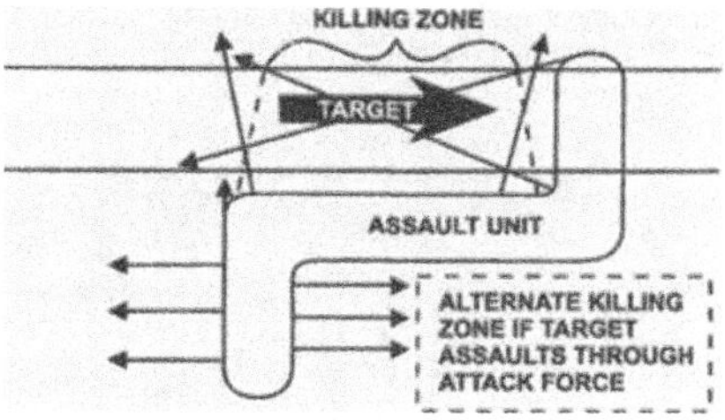

Figure D-6. "Z" Formation.

The additional wing may serve any of the following purposes:

- To engage an enemy force attempting to relieve or reinforce the enemy unit engaged in the kill zone.
- To seal the end of the killing zone.
- To restrict a flank.
- To prevent an envelopment.

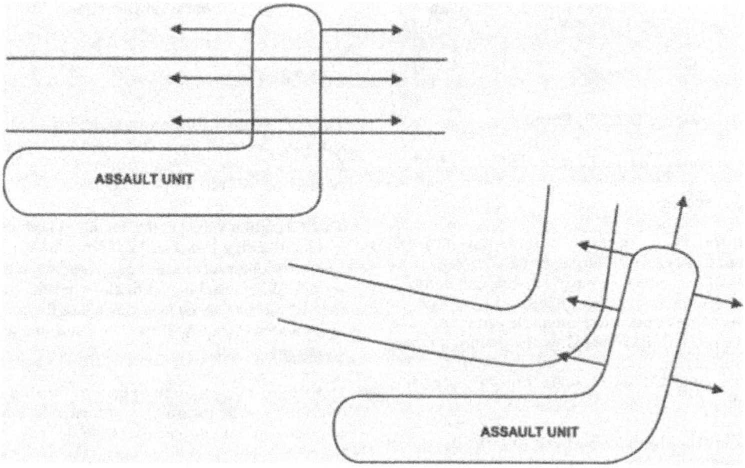

Figure D-5. "L" Formation.

"T" Formation

In the "T" shaped formation, the assault element is deployed across (perpendicular to) the enemy's route of movement so that its position forms the crossing of a "T" at the top. (See fig. D-7.)

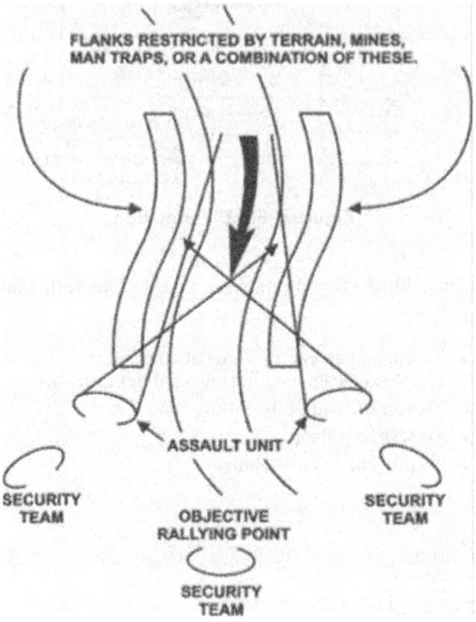

Figure D-7. "T" Formation.

This formation can be used day or night to establish an ambush to interdict movement through open areas that are hard to seal off.

A small force can use the "T" formation to harass, slow, and disorganize a larger force. When the lead elements of the enemy are engaged, they will normally attempt to maneuver right or left to close with the ambush. Mines and other obstacles placed to the flanks of the killing zones slow the enemy's movements and permit the ambush force to deliver heavy fire and withdraw without becoming decisively engaged.

The "T" formation can be used to interdict small groups attempting night movement across open areas. For example, the assault element is deployed along an avenue of approach with every second man facing the opposite direction. The attack of the enemy approaching from either direction requires only that every second man may shift to the opposite side of the formation. Each man fires only to his front and only when the enemy is at a very close range. Attack is by fire only and each man keeps the enemy under fire as long as it remains to his front.

If the enemy attempts to escape in either direction along the killing zone, each Marine takes the enemy under fire as the enemy comes into the Marine's sector of fire. The "T" formation is very effective at halting infiltration. But it has one chief disadvantage; there is a possibility that the ambush will engage a superior force at night while spread out. (See fig. D-8.)

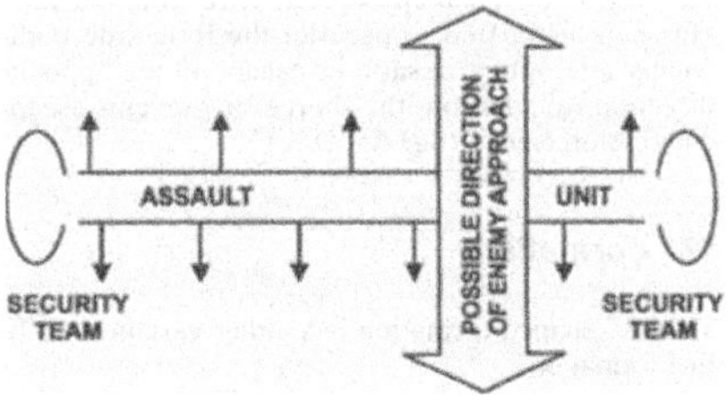

Figure D-8. "T" Formation (Target Approaching from Either Direction).

The "V" Formation

The "V" shaped formation is deployed along both sides of the enemy's route of movement so that it forms a "V"; care is taken to ensure that neither group (within the "V") fires into the other.

This formation subjects the enemy to both enfilading and interlocking fire. The "V" formation is best suited for fairly open terrain but can also be used in close terrain. When established in close terrain, the legs of the "V" close in as the lead element of the enemy force approaches the apex of the "V", and opens fire at a close range.

Here, even more than in open terrain, all movement and fire must be carefully coordinated and controlled to ensure that the fire of one wing does not endanger the other wing. The wider separation of forces makes this formation difficult to control, and there are few sites that favor its use. Its main advantage is that it is difficult for the enemy to detect the ambush until it is well into the killing zone. (See figs. D-9 and D-10.)

Triangle Formation

The triangle is a variation of the "V" formation and can be varied in three ways:

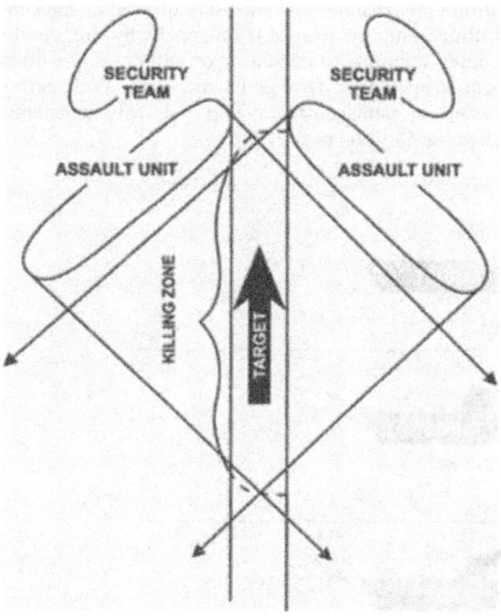

Figure D-9. "V" Formation (Open Terrain).

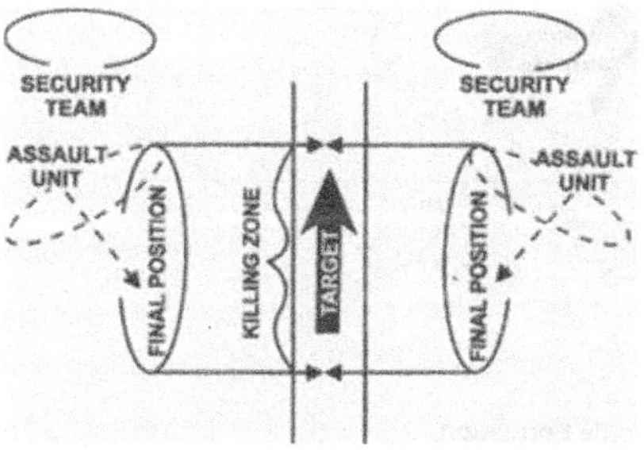

Figure D-10. "V" Formation (Close Terrain).

Closed Triangle Formation

(See fig. D-11.) The assault element is deployed in three teams, positioned so that they form a triangle (or closed "V"). An automatic weapon is placed at each point of the triangle and positioned so that it can be shifted quickly to interlock with either of the others. Men are positioned so that their sectors of fire overlap. Mortars may be positioned inside the triangle.

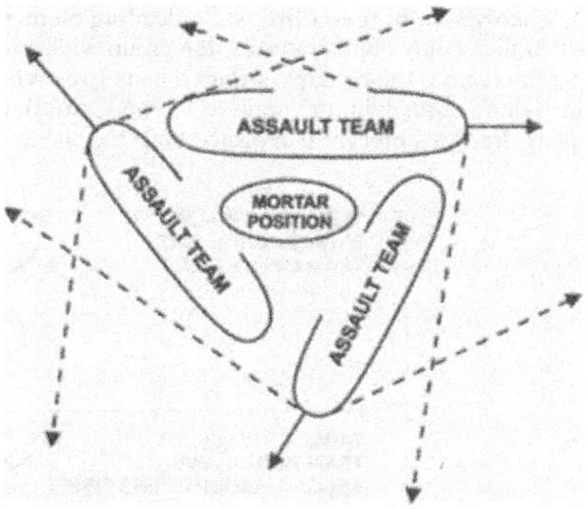

Figure D-11. Closed Triangle Formation (Night Ambush).

When deployed in this manner, the triangle ambush is used to interdict night movement through open areas. When enemy approach is likely to be from any direction, this formation provides all-around security, and security forces are deployed only when they can be positioned so that if detected by an approaching enemy, they will not compromise the ambush. Attack is by fire only, and the enemy is allowed to approach within close range before fire is initiated.

The advantages of the closed triangle formation are ease of control, all-around security, and the enemy can be brought under the fire of at least two automatic weapons, regardless of the direction they approach.

Disadvantages are that it requires a force of platoon size or larger to reduce the danger of being overrun by an unexpectedly large force and that one or more sides of the triangle may come under enfilade fire. The lack of dispersion, particularly at the points, increases the danger from enemy mortar fire.

Open Triangle Harassing Formation

This variation of the triangle ambush enables a small force to harass, slow, and inflict heavy casualties upon a large force without itself being decisively engaged. The assault element is deployed in three teams, positioned so that each team becomes a corner of a triangle containing the killing zone. (See fig. D-12.)

When the enemy enters the killing zone, the team to the enemy's front opens fire on the leading element. When the enemy counterattacks, the group withdraws and the team on the enemy's flank opens fire. When this team is attacked, the team to the opposite flank opens fire. This process is repeated until the enemy is pulled apart. Each team reoccupies its position, if possible, and continues to inflict the maximum damage possible without becoming decisively engaged.

Open Triangle Destruction Formation

The assault element is again deployed in three teams, positioned so that each team is a point of the triangle, 200 to 300 meters apart. The killing zone is the area within the triangle. The enemy is allowed to enter the killing zone; the nearest team attacks by fire. As the enemy attempts to maneuver or withdraw, the other teams open fire. One or more teams, as directed, assault or maneuver to envelop or destroy the enemy. (See fig. D-13 on page D-6.)

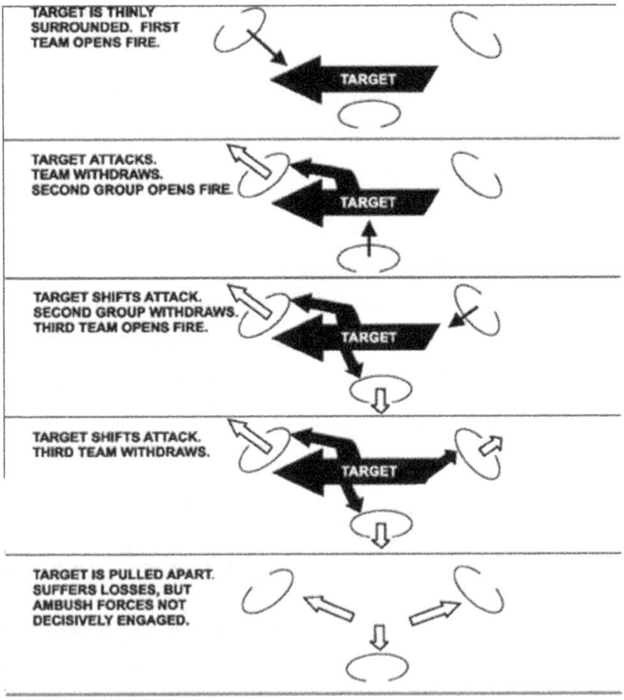

Figure D-12. Open Triangle Formation.

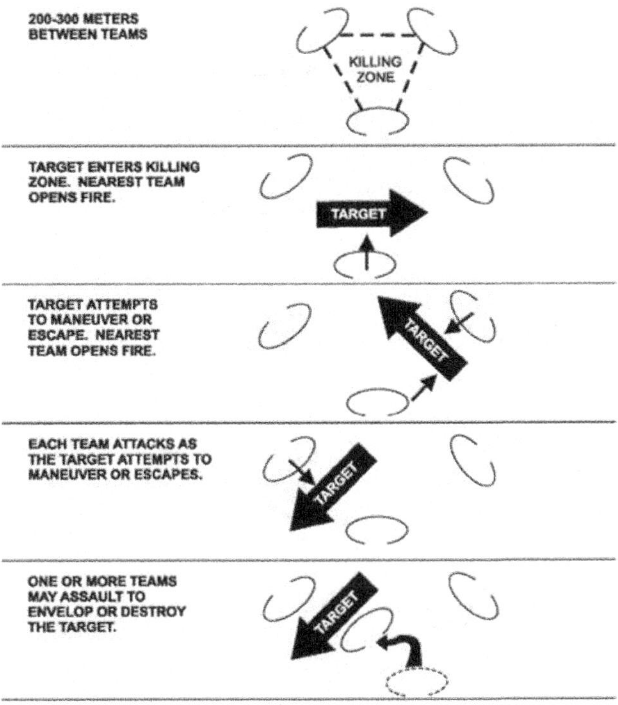

Figure D-13. Open Triangle Formation (Destruction Ambush).

This formation is suitable for platoon-size or larger ambush forces. A smaller force would be in too great a danger of being overrun. Another disadvantage is that control, in assaulting or maneuvering, is very difficult. Very close coordination and control are necessary to ensure that assaulting or maneuvering teams are not fired on by another team. The ambush site must be a fairly level open area that provides concealment around its border for the ambush force.

The Box Formation

The "box" formation is similar in purpose to the open triangle ambush. The assault element is deployed in four teams, positioned so that each team becomes a corner of a square or rectangle containing the killing zone. It can be used for a harassing or destruction ambush in the same manner as the variations of the open triangle formation. (See figs D-14 on page D-7 and D-15 on page D-8.)

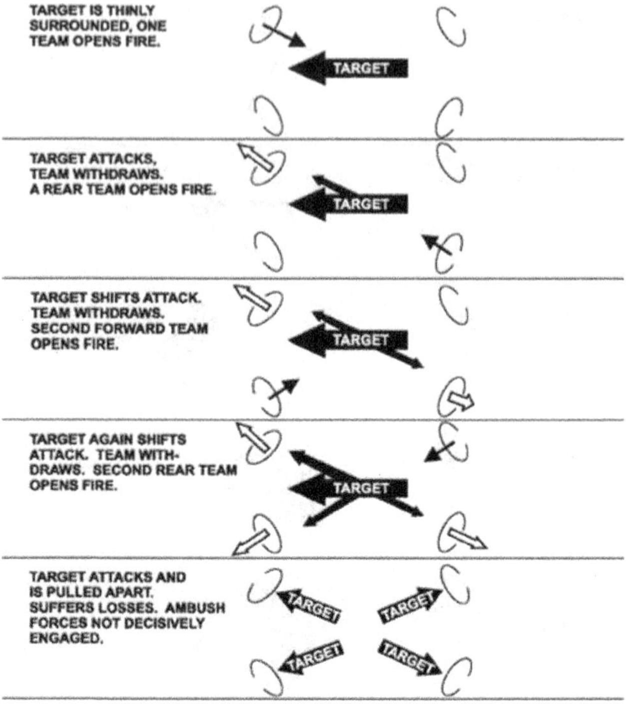

Figure D-14. Box Formation (Harrassing Ambush).

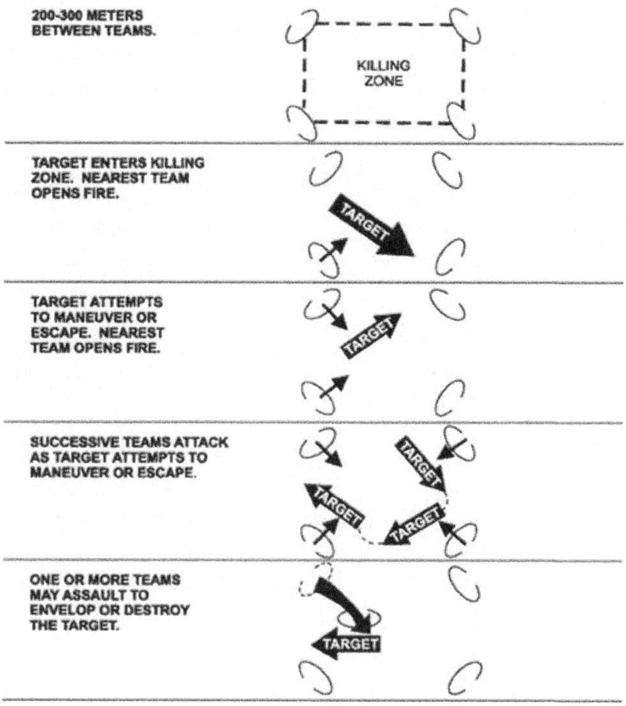

Figure D-15. Box Formation (Destruction Ambush).

Appendix E. Acronyms

A&S . . . assault and security

CIT . . . counterintelligence team

EOD . . . explosive ordnance disposal

ITT . . . interrogator-translator team

MAGTF . . . Marine air-ground task force
MEDEVAC . . . medical evacuation
METT-T . . . mission, enemy, terrain and weather, troops and support available

NBC . . . nuclear, biological, chemical
NCO . . . noncommissioned officer

OTL . . . overserver-target line

RCA . . . riot control agents
ROE . . . rules of engagement

SMAW . . . shoulder-launched multipurpose assault weapon
SMEAC . . . situation, mission, execution, administration and logistics, and command and signal
SOP . . . standing operating procedures
SPOTREP . . . spot report

TTP . . . tactics, techniques, and procedures

Appendix F. References

Joint Publication (JP)

1-02 DOD Dictionary of Military and Associated Terms

Fleet Marine Force Manual (FMFM)

6-5 Marine Rifle Squad (proposed MCWP 3-11.2)

Marine Corps Warfighting Publications (MCWP)

3-15.3 Scout Sniping (under development)

3-35.3 Military Operations on Urbanized Terrain (MOUT)

3-41.2 Raids (under development)

Marine Corps Reference Publications (MCRP)

2-15.3A Reconnaissance Patrol Leader's Planning Handbook (under development)

2-15.3B Reconnaissance Reports Guide

3-02C Water Survival Handbook (under development, currently exists as Fleet Marine Force Manual 0-13)

5-12A Operational Terms and Graphics

5-12C Marine Corps Supplement to the DOD Dictionary of Military and Associated Terms

Made in the USA
Monee, IL
05 November 2026

67811378R00118